RAND McNALLY

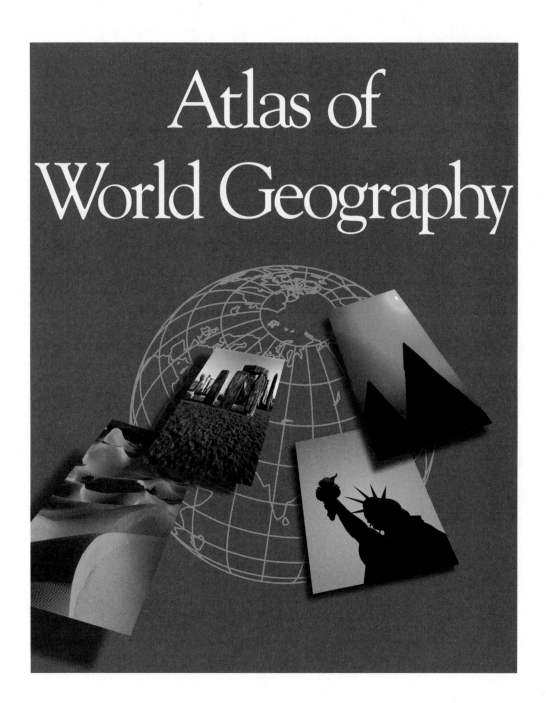

Atlas of World Geography

Editors
Brett R. Gover
Margaret McNamara
Ann T. Natunewicz

Design
Donna M. McGrath
Peggy Hogan

Cover Design
Brian C. Doherty

Design Production
DePinto Graphic Design

Cartographic Direction
V. Patrick Healy

Cartographic Staff
Nina Lusterman
Robert K. Argersinger
Lynn Jasmer
Patty A. Porter
James A. Purvis
L. Charlene Smith
James A. Wooden
David C. Zapenski

Research
Susan K. Hudson

Continental Thematic Maps
Thomas F. Vitacco
Marzee L. Eckhoff
Gwynn A. Lloyd
Robert L. Merrill
David R. Simmons
Ashley James Snyder
Barbara Benstead-Strassheim
Dara L. Thompson

Photo Research
Feldman and Associates, Inc.

Photograph Credits:

Front cover images provided by © 1997 PhotoDisc, Inc.

H. Armstrong Roberts: © R. Kord, 7, 31 (b); © Zefa, 100; © M. Schneiders, 108; © Smith/Zefa, 150 (b)

© Randall Hyman: 101

© PhotoDisc, Inc.: 135

Odyssey Chicago: © R. Frerck, 152

Tony Stone Images: © K. Wood, 29 (t); © Mark Segal, 68; © David Frazier, 69; © Jacques Jangoux, 98 (t); © John Warden, 98 (b); © Owen Franken, 110; © John Lamb, 122 (t); © Yann Layma, 134; © Paul Chelsey, 153

Information Credits:

Volcano data, pages 29, 32, and 36: Tom Simkin, Smithsonian Institution Global Volcanism Program

Earthquake data, page 37: Paula Dunbar, National Geophysical Data Center, National Oceanic and Atmospheric Administration

Australia information, page 38: Australian Tourist Commission

Much of the information on the destruction of the Amazonian rain forest, page 101, was provided by Fred Engel of the Center for Earth and Planetary Science, National Air and Space Museum, Smithsonian Institution, Washington, D.C.

Atlas of World Geography

Published and printed in the United States of America

ISBN: 0-528-17790-0

For information about ordering *Atlas of World Geography,* call 1-800-678-7263; or visit our website at:

www.K12online.com

Table of Contents

Using the Atlas

Maps and Atlases

Today, satellite images (Figure 1) and aerial photography show us the face of the Earth in precise detail. It is hard to imagine how difficult it once was to ascertain what our planet looked like—even small parts of it. Yet from earliest history we have evidence of humans trying to depict the world through maps and charts.

Figure 1

Twenty-five hundred years ago, on a tiny clay tablet the size of a hand, the Babylonians inscribed the earth as a flat disk (Figure 2) with Babylon at the center. The section of the Cantino map of 1502 (Figure 3) is an example of a portolan chart used by mariners to chart the newly discovered Americas. Handsome and useful maps have been produced by many cultures. The Mexican map (Figure 4) drawn in 1583 marks hills with wavy lines and roads with footprints between parallel lines. The methods and materials used to create these maps were dependent upon the technology available, and their accuracy suffered considerably. The maps in this atlas show the detail and accuracy that cartographers are now able to achieve. They benefit from our ever-increasing technology, including satellite imagery and computer-assisted cartography.

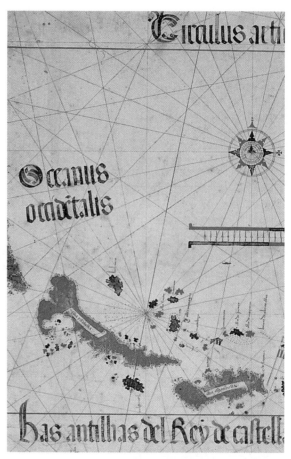

Figure 3

Figure 4

Figure 2

In 1589, Gerardus Mercator used the word "atlas" to describe a collection of maps. Atlases have become a unique and indispensable reference for graphically defining the world and answering the question "Where?" Only on a map can the countries, cities, roads, rivers, and lakes covering a vast area be simultaneously viewed in their relative locations. Routes between places can be traced, trips planned, boundaries of neighboring states and countries examined, distances between places measured, the meandering of rivers and streams and the sizes of lakes visualized, and remote places imagined.

Getting the Information

An atlas can be used for many purposes, from planning a trip to finding hot spots in the news and supplementing world knowledge. To realize the potential of an atlas, the user must be able to:

1) Find places on the maps
2) Measure distances
3) Determine directions
4) Understand map symbols

Finding Places

One of the most common and important tasks facilitated by an atlas is finding the location of a place in the world. A river's name in a book, a city mentioned in the news, or a potential vacation spot may prompt your need to know where the place is located. The illustrations and text below explain how to find Lagos, Nigeria.

1) Look up the place-name in the index at the back of the atlas. Lagos, Nigeria can be found in the map on page 128, and it can be located on the map by its latitude and longitude, expressed in degrees: 7 North Latitude, 3 East Longitude (Figure 5).

La Fayette, In., U.S. 40N	87W	**90**
Lafayette, La., U.S. 30N	92W	**95**
Laghouat, Alg. 34N	3 E	**128**
Lagos, Nig. 7N	3 E	**128**
La Grande, Or., U.S. 45N	118W	**82**
LaGrange, Ga., U.S. 33N	85W	**92**
Lahore, Pak. 32N	74 E	**143**
Lahti, Fin. 61N	26 E	**116**

Figure 5

2) Turn to the map of Northern Africa on page 128. Note that the latitude appears in the right and left margins of the map, and the longitude in the upper and lower margins.

3) To find Lagos on the map, place your left index finger on the left margin at 7 degrees (between 5 and 10); and your right index finger in the top margin at 3 degrees East (between 0 and 5). Move your left finger across the map and your right finger down the map. Your fingers will meet in the area in which Lagos is located (Figure 6).

Figure 6

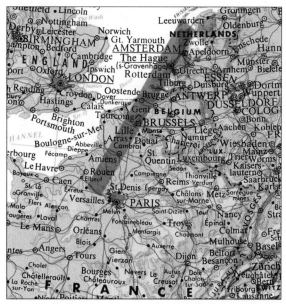

Figure 7

Measuring Distances

In planning trips, determining the distances between two places is essential, and an atlas can help in travel preparation. For instance, to determine the approximate distance between Paris, France and Amsterdam, Netherlands, follow these three steps:

1) Lay a slip of paper on the map on page 117 so that its edge touches the two cities. Adjust the paper so only one corner touches Paris. Mark the paper directly at the spot where Amsterdam is located (Figure 7).

2) Place the paper along the scale of miles beneath the map. Position the corner at 0 and line up the edge of the paper along the scale. The pencil mark on the paper indicates Amsterdam is between 250 and 300 miles from Paris (Figure 8).

3) To find the exact distance, make a second pencil mark at the 250-mile point of the scale. Then slide the paper to the left so that this second mark is lined up with 0 on the scale (Figure 9). The Amsterdam mark now falls at the third 10-mile point on the scale. This means that the Paris and Amsterdam are approximately 250 plus 30—or 280—miles apart.

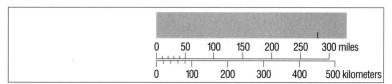

Figure 8

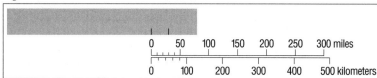

Figure 9

Determining Directions

Most of the maps in the atlas are drawn so that when oriented for normal reading, north is at the top of the map, south is at the bottom, west is at the left, and east is at the right. Most maps have a series of lines drawn across them—the lines of latitude and longitude. Lines of latitude, or parallels of latitude, are drawn east and west. Lines of longitude, or meridians of longitude, are drawn north and south (Figure 10, at bottom of page).

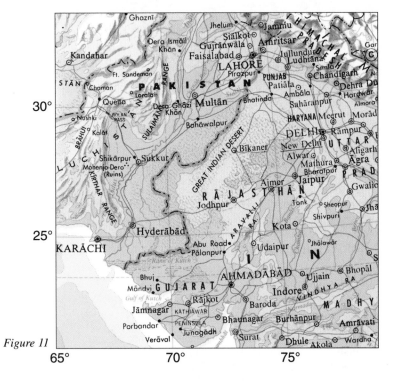

Figure 11

Parallels and meridians appear as either curved or straight lines. For example, in the section of the map of Southwestern Asia (Figure 11), from page 143, the parallels of latitude appear as curved lines. The meridians of longitude are curved vertical lines.

Latitude and longitude lines help locate places on maps. Parallels of latitude are numbered in degrees north and south of the Equator. Meridians of longitude are numbered in degrees east and west of a line called the Prime Meridian, running through Greenwich, England, near London. Any place on Earth can be located by the latitude and longitude lines running through it.

To determine directions or locations on the map, you must use the parallels and meridians. For example, suppose you want to know which is farther north, Karachi, Pakistan or Delhi, India. The map in Figure 11 shows that Karachi is south of the 25° parallel of latitude and that Delhi is north of it. Therefore Delhi is farther north than Karachi. By looking at the meridians of longitude, you can determine which city is farther east. Karachi is approximately 2° east of the 65° meridian, and Delhi is about 2° east of the 75° meridian. Delhi is farther east than Karachi.

Understanding Map Symbols

In a very real sense, every map is a symbol representing the world or part of it. It is a reduced representation of the Earth: each of the world's features —cities, rivers, etc.—is represented by a symbol. Map symbols may take the form of points, such as dots or squares (often used for cities, capital cities, or points of interest) or lines (roads, railroads, rivers). Symbols may also occupy an area, showing extent of coverage (terrain, forests, deserts). They seldom look like the feature they represent and therefore must be identified and interpreted. For instance, some of the maps in this atlas define political units by a colored line depicting their boundaries. Neither the colors nor the boundary lines are actually found on the surface of the Earth, but because countries and states are such important political components of the world, strong symbols are used to represent them. The Legend on page 51 of this atlas identifies the symbols used on the maps.

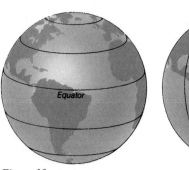
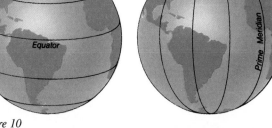
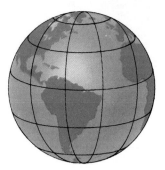

Figure 10

Introduction to Geographic Tables, Charts, and Graphs

This section provides an interesting way to learn key geographic information about your country and the world. Locate places referred to in the questions on Atlas of World Geography maps.

The Universe and Solar System

The Milky Way Galaxy

Our star, the Sun, is one of 200 billion stars banded together in the enormous gravitational spiral nebula called the Milky Way Galaxy, which is but one of millions of known galaxies in the universe.

The Milky Way is huge; it would take light — which travels at 186,000 miles per second — 100,000 years to go from one end of the galaxy to the other. In addition to the billions of stars, Earth shares the Milky Way with eight other known planets.

Statistical Data for the Milky Way Galaxy

Diameter: 100,000 light-years

Mass: About 200 billion suns

Distance between spiral arms: 6,500 light years

Thickness of galactic disk: 1,300 light-years

Satellite galaxies: 2 (visible only in the southern sky)

Sun

The Sun's diameter — more than 865,000 miles — is 109 times greater than that of the Earth. Even so, the Sun is actually a fairly small star. Somewhere in the vastness of the universe astronomers have located a star that is 3,500 times larger than the Sun.

Diameter: 865,000 miles (1,392,000 km)
Mass: 333,000 times that of the Earth
Surface temperature: 10,300° F (5,700° C)
Central temperature: 27 million° F (15 million° C)
Composition: 70% hydrogen, 27% helium
Spin (at equator): 26 days, 21 hours

Mercury

Distance from the Sun: 35,985,000 miles (57,909,000 km), or 39% that of the Earth
Diameter: 3,031 miles (4,878 km), or 38% that of the Earth
Average surface temperature: 340° F (171° C)
Atmosphere: Extremely thin, contains helium and hydrogen
Length of day: 58 days, 15 hours, 30 minutes
Length of year: 87.97 days
Satellites: None

Venus

Distance from the Sun: 67,241,000 miles (108,209,000 km), or 72% that of the Earth
Diameter: 7,521 miles (12,104 km), or 95% that of the Earth
Surface temperature: 867° F (464° C)
Surface pressure: 90 times that of the Earth, equivalent to the pressure at a water depth of 3,000 feet (900 meters)
Atmosphere: 96% carbon dioxide
Length of day: 243 days, 14 minutes. The planet spins opposite to the rotation of the Earth.
Length of year: 224.7 days
Satellites: None

Earth

Distance from the Sun: 92,960,000 miles (149,598,000 km)
Diameter: 7,926 miles (12,756 km)
Average surface temperature: 58° F (14° C)
Surface pressure: 1 atmosphere
Atmosphere: 78% nitrogen, 21% oxygen
Length of day: 23 hours, 56 minutes and 4 seconds
Length of year: 365.25 days
Satellites: 1

The Moon

The Moon is the Earth's only natural satellite. About 2,160 miles (3,746 km) across, the Moon is an airless, waterless world just one-fourth the size of the Earth. It circles the planet once every 27 days at an average distance of about 238,000 miles (384,000 km).

Jupiter

By any measure, Jupiter is the solar system's giant. To equal Jupiter's bulk would take 318 Earths. Over 1,300 Earth-sized balls could fit within this enormous planet.

Distance from the Sun: 483,631,000 miles (778,292,000 km), or 5.2 times that of the Earth
Diameter: 88,700 miles (142,800 km), or 11.3 times that of the Earth
Temperature at cloud tops: −234° F (−148° C)

Mars

Distance from the Sun: 141,642,000 miles (227,940,000 km), about 1.5 times that of the Earth
Diameter: 4,222 miles (6,794 km), or 53% that of the Earth
Average surface temperature: −13° F (−25° C)
Surface pressure: 0.7% (1/150 th) that of the Earth
Atmosphere: 95% carbon dioxide, 2.7% nitrogen
Length of day: 24 hours, 37 minutes
Length of year: 1 year, 321.73 days
Satellites: 2

Spatial Relationships of the Sun and the Planets

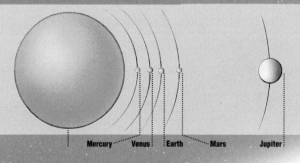

Mercury   Venus   Earth   Mars   Jupiter   Saturn

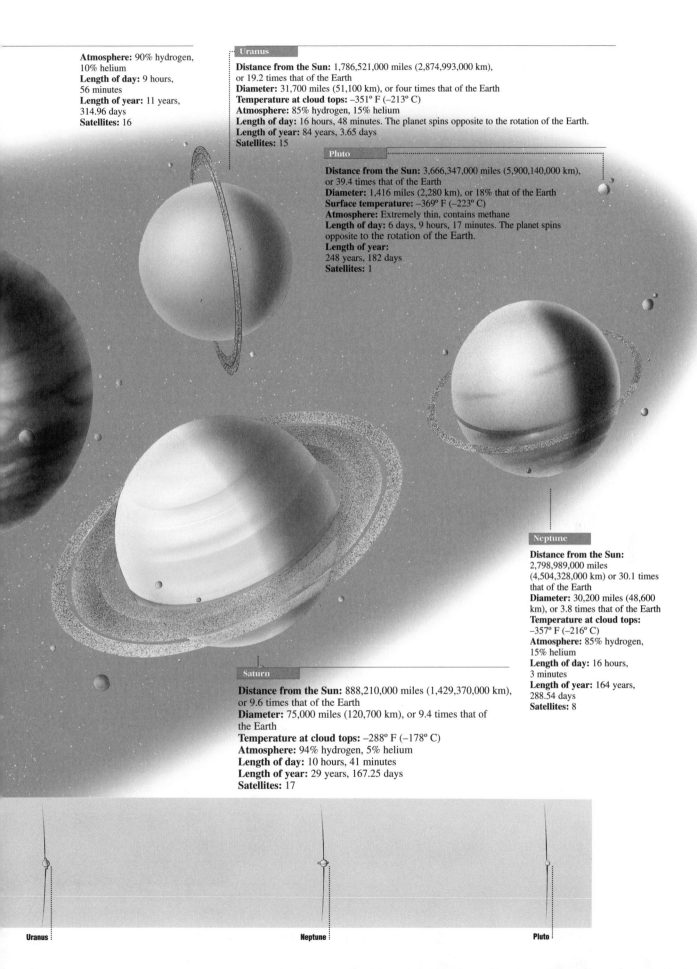

Atmosphere: 90% hydrogen, 10% helium
Length of day: 9 hours, 56 minutes
Length of year: 11 years, 314.96 days
Satellites: 16

Uranus

Distance from the Sun: 1,786,521,000 miles (2,874,993,000 km), or 19.2 times that of the Earth
Diameter: 31,700 miles (51,100 km), or four times that of the Earth
Temperature at cloud tops: –351° F (–213° C)
Atmosphere: 85% hydrogen, 15% helium
Length of day: 16 hours, 48 minutes. The planet spins opposite to the rotation of the Earth.
Length of year: 84 years, 3.65 days
Satellites: 15

Pluto

Distance from the Sun: 3,666,347,000 miles (5,900,140,000 km), or 39.4 times that of the Earth
Diameter: 1,416 miles (2,280 km), or 18% that of the Earth
Surface temperature: –369° F (–223° C)
Atmosphere: Extremely thin, contains methane
Length of day: 6 days, 9 hours, 17 minutes. The planet spins opposite to the rotation of the Earth.
Length of year: 248 years, 182 days
Satellites: 1

Neptune

Distance from the Sun: 2,798,989,000 miles (4,504,328,000 km) or 30.1 times that of the Earth
Diameter: 30,200 miles (48,600 km), or 3.8 times that of the Earth
Temperature at cloud tops: –357° F (–216° C)
Atmosphere: 85% hydrogen, 15% helium
Length of day: 16 hours, 3 minutes
Length of year: 164 years, 288.54 days
Satellites: 8

Saturn

Distance from the Sun: 888,210,000 miles (1,429,370,000 km), or 9.6 times that of the Earth
Diameter: 75,000 miles (120,700 km), or 9.4 times that of the Earth
Temperature at cloud tops: –288° F (–178° C)
Atmosphere: 94% hydrogen, 5% helium
Length of day: 10 hours, 41 minutes
Length of year: 29 years, 167.25 days
Satellites: 17

Uranus   Neptune   Pluto

The Earth

History of the Earth

Estimated age of the Earth:
At least 4.6 billion (4,600,000,000) years.

Formation of the Earth:
It is generally thought that the Earth was formed from a cloud of gas and dust (A) revolving around the early Sun. Gravitational forces pulled the cloud's particles together into an ever denser mass (B), with heavier particles sinking to the center. Heat from radioactive elements caused the materials of the embryonic Earth to melt and gradually settle into core and mantle layers. As the surface cooled, a crust formed. Volcanic activity released vast amounts of steam, carbon dioxide and other gases from the Earth's interior. The steam condensed into water to form the oceans, and the gases, prevented by gravity from escaping, formed the beginnings of the atmosphere (C).

The calm appearance of our planet today (D) belies the intense heat of its interior and the violent tectonic forces which are constantly reshaping its surface.

Ⓐ

Ⓑ

Ⓒ

Ⓓ

Periods in Earth's history

Earth's history is divided into different **eras**, which are subdivided into **periods**.

The most recent periods are themselves subdivided into **epochs**. The main divisions and subdivisions are shown below.

	Began	Ended	
	(million years ago)		
Precambrian Era			
Archean Period	3,800	2,500	Start of life
Proterozoic Period	2,500	590	Life in the seas
Paleozoic Era			
Cambrian Period	590	500	Sea life
Ordovician Period	505	438	First fishes
Silurian Period	438	408	First land plants
Devonian Period	408	360	Amphibians
Carboniferous Period	360	286	First reptiles
Permian Period	286	248	Spread of reptiles
Mesozoic Era			
Triassic Period	248	213	Reptiles and early mammals
Jurassic Period	213	144	Dinosaurs
Cretaceous Period	144	65	Dinosaurs, dying out at the end
Cenozoic Era			
Tertiary Period			
Paleocene	65	55	Large mammals
Eocene	55	38	Primates begin
Oligocene	38	25	Development of primates
Miocene	25	5	Modern-type animals
Pliocene	5	2	*Australopithecus* ape, ancestor to the human race
Quaternary Period			
Pleistocene	2	0.01	Ice ages; true humans
Holocene	0.01	Present	Modern humans

Source: *Atlas of the Universe* by Patrick Moore, Reed International Books Limited, 1994.

Internal Structure of the Earth

In its simplest form, the Earth is composed of a crust, a mantle with an upper and lower layer, and a core, which has an inner region.

Temperatures in the Earth increase with depth, as is observed in a deep mine shaft or borehole, but the prediction of temperatures within the Earth is made difficult by the fact that different rocks conduct heat at different rates: rock salt, for example, has 10 times the heat conductivity of coal. Also, estimates have to take into account the abundance of heat-generating atoms in a rock. Radioactive atoms are concentrated toward the Earth's surface, so the planet has, in effect, a thermal blanket to keep it warm. The temperature at the center of the Earth is believed to be approximately 5,400° F (3,000° C).

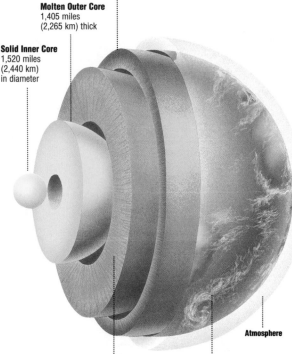

Upper Mantle
415 miles
(667 km) thick

Molten Outer Core
1,405 miles
(2,265 km) thick

Solid Inner Core
1,520 miles
(2,440 km)
in diameter

Lower Mantle
1,365 miles
(2,200 km) thick

Solid Crust
0–19 miles
(0–33 km) thick

Atmosphere

Chemical composition of the Earth:

The chemical composition of the Earth varies from crust to core. The upper crust of continents, called sial, is mainly granite, rich in aluminum and silicon. Oceanic crust, or sima, is largely basalt, made of magnesium and silicon. The mantle is composed of rocks that are rich in magnesium and iron silicates, whereas the core, it is believed, is made of iron and nickel oxides.

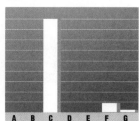

- Sial
- Sima
- Upper Mantle
- Lower Mantle
- Outer Core
- Inner Core

A. Silicon
B. Aluminum
C. Iron
D. Calcium
E. Magnesium
F. Nickel
G. Other

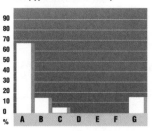

Sial (upper crust of continents)

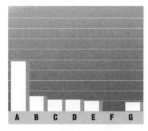

Sima (oceanic crust)

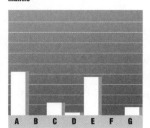

Mantle

Core

Measurements of the Earth

Equatorial circumference of the Earth: 24,901.45 miles (40,066.43 km)

Polar circumference of the Earth: 24,855.33 miles (39,992.22 km)

Equatorial diameter of the Earth: 7,926.38 miles (12,753.54 km)

Polar diameter of the Earth: 7,899.80 miles (12,710.77 km)

Equatorial radius of the Earth: 3,963.19 miles (6,376.77 km)

Polar radius of the Earth: 3,949.90 miles (6,355.38 km)

Estimated weight of the Earth:
6,600,000,000,000,000,000,000 tons, or 6,600 billion billion tons (5,940 billion billion metric tons)

Total surface area of the Earth: 197,000,000 square miles (510,230,000 sq km)

Total land area of the Earth (including inland water and Antarctica): 57,900,000 square miles (150,100,000 sq km)

Total ocean area of the Earth: 139,200,000 square miles (360,528,000 sq km), or 70% of the Earth's surface area

Total area of the Earth's surface covered with water (oceans and all inland water): 147,750,000 square miles (382,672,500 sq km), or 75% of the Earth's surface area

Types of water: 97% of the Earth's water is salt water; 3% is fresh water

Life on Earth

Number of plant species on Earth: About 350,000

Number of animal species on Earth: More than one million

Estimated total human population of the Earth: 6,195,885,000

Movements of the Earth

Mean distance of the Earth from the Sun: About 93 million miles (149.6 million km)

Period in which the Earth makes one complete orbit around the Sun: 365 days, 5 hours, 48 minutes, and 46 seconds

Speed of the Earth as it orbits the Sun: 66,700 miles (107,320 km) per hour

Period in which the Earth makes one complete rotation on its axis: 23 hours, 56 minutes and 4 seconds

Equatorial speed at which the Earth rotates on its axis: More than 1,000 miles (1,600 km) per hour

The Shape of the Earth

Comparing the Earth's equatorial and polar dimensions reveals that our planet is actually not a perfect sphere but rather an oblate spheroid, flattened at the poles and bulging at the equator. This is the result of a combination of gravitational and centrifugal forces.

An even more precise term for the Earth's shape is "geoid" — the actual shape of sea level, which is lumpy, with variations away from spheroid of up to 260 feet (80 m). This lumpiness reflects major variations in density in the Earth's outer layers.

The Seasons
(Northern Hemisphere)

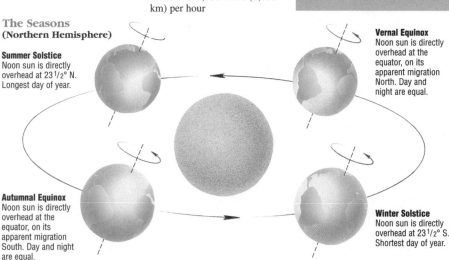

Summer Solstice
Noon sun is directly overhead at 23 1/2° N. Longest day of year.

Vernal Equinox
Noon sun is directly overhead at the equator, on its apparent migration North. Day and night are equal.

Autumnal Equinox
Noon sun is directly overhead at the equator, on its apparent migration South. Day and night are equal.

Winter Solstice
Noon sun is directly overhead at 23 1/2° S. Shortest day of year.

Plate Tectonics

Continental Drift

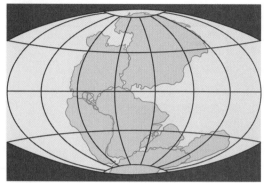

225 million years ago the supercontinent of Pangaea exists and Panthalassa forms the ancestral ocean. Tethys Sea separates Eurasia and Africa.

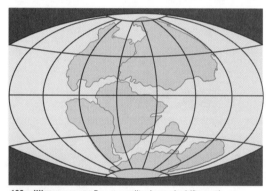

180 million years ago Pangaea splits, Laurasia drifts north. Gondwanaland breaks into South America/Africa, India, and Australia/Antarctica.

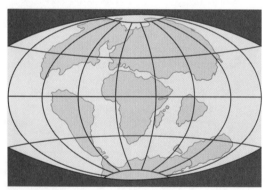

65 million years ago ocean basins take shape as South America and India move from Africa and the Tethys Sea closes to form the Mediterranean Sea.

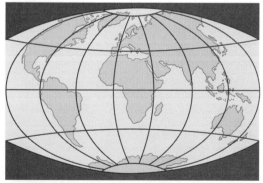

The present day: India has merged with Asia, Australia is free of Antarctica, and North America is free of Eurasia.

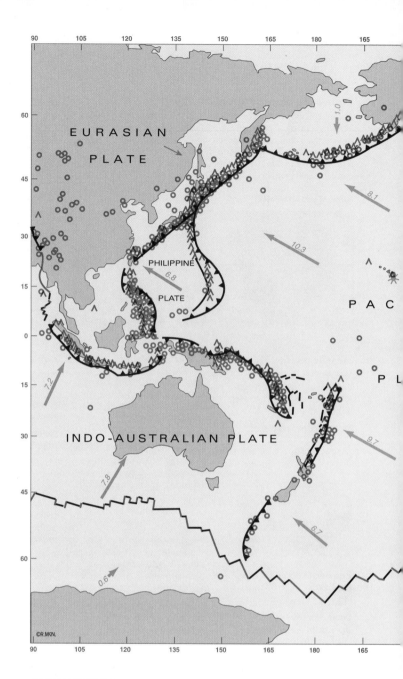

PLATE TECTONICS

Types of plate boundaries

Divergent: magma emerges from the earth's mantle at the mid-ocean ridges forming new crust and forcing the plates to spread apart at the ridges.

Convergent: plates collide at subduction zones where the denser plate is forced back into the earth's mantle forming deep ocean trenches.

Transform: plates slide past one another producing faults and fracture zones.

Other map symbols

→ Direction of plate movement

6.7 → Length of arrow is proportional to the amount of plate movement (number indicates centimeters of movement per year)

○ Earthquake of magnitude 7.5 and above (from 10 A.D. to the present)

Λ Volcano (eruption since 1900)

✳ Selected hot spots

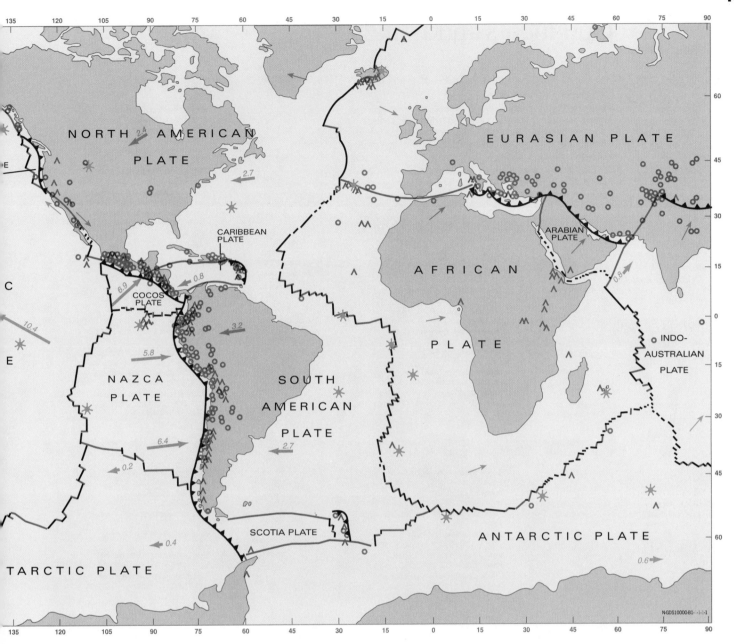

According to plate tectonic theory, Earth's lithosphere—the crust and uppermost rigid part of the mantle—is divided into plates that move relative to one another. The map above shows the locations and names of the major plates.

These rigid plates move on a molten layer of Earth's mantle called the asthenosphere. The moving plates meet at three different types of boundaries—divergent, convergent, and transform. (See map legend at the left.)

Divergent Boundaries

At divergent boundaries, plates are pushed apart by currents in the asthenosphere in a process called rifting. Most rifting occurs on the ocean floors. In ocean floor rifting, molten material from the asthenosphere wells up between the separated plates, hardens, and forms ridges. (See World Physical Map and Ocean Depths Profile, pages 54 and 55.) This process of adding to the Earth's crust is called seafloor spreading.

Convergent Boundaries

At convergent boundaries, the moving plates collide with one another. The edge of the heavier plate sinks under the crust of the lighter plate, and is consumed back into Earth's mantle in a process called subduction. Subduction can create deep ocean trenches as the crust

of the lighter plate sinks into the mantle. (See World Physical Map, pages 54 and 55.) The colliding plates also create mountain chains as the lighter plate is pushed up over the heavier plate.

Transform Boundaries

When plates meet at transform boundaries or faults, they grind past each other. This movement neither increases nor destroys Earth's crust. The San Andreas Fault, north of San Francisco, is a famous example of a transform boundary.

Earthquakes and Volcanoes

Most of Earth's volcanoes and earthquakes occur along plate boundaries. The ring of volcanic and seismic activity along the west coasts of North and South America and the east coast of Asia, known as the Ring of Fire, follows plate boundaries. Volcanoes and earthquakes also occur at locations known as hot spots, where hot rock from deep in the mantle rises to the surface, creating some of Earth's tallest mountains.

Continental Drift

Plate tectonic theory assumes that the rigid plates have moved slowly through the millennia, carrying the continents with them. The history of this continental drifting is illustrated by the four maps to the left.

Continents and Islands

The word "continents" designates the largest continuous masses of land in the world.

For reasons that are mainly historical, seven continents are generally recognized: Africa, Antarctica, Asia, Australia, Europe, North America, and South America. Since Asia and Europe actually share the same land mass, they are sometimes identified as a single continent, Eurasia.

The lands of the central and south Pacific, including Australia, New Zealand, Micronesia, Melanesia, and Polynesia, are sometimes grouped together as Oceania.

The Continents

Africa

Area in square miles (sq km):
11,700,000 (30,300,000)
Estimated population:
832,590,000
Population per square mile (sq km):
71 (27)
Mean elevation in feet (meters):
1,900 (580)
Highest elevation in feet (meters):
Kilimanjaro, Tanzania, 19,340 (5,895)
Lowest elevation in feet (meters):
Lac Assal, Djibouti, 515 (157) below sea level

Antarctica

Area in square miles (sq km):
5,400,000 (14,000,000)
Estimated population:
Uninhabited
Population per square mile (sq km):
0 (0)
Mean elevation in feet (meters):
6,000 (1,830)
Highest elevation in feet (meters):
Vinson Massif, 16,066 (4,897)
Lowest elevation in feet (meters):
Deep Lake, 184 (56) below sea level

Asia

Area in square miles (sq km):
17,300,000 (44,900,000)
Estimated population:
3,761,165,000
Population per square mile (sq km):
217 (84)
Mean elevation in feet (meters):
3,000 (910)
Highest elevation in feet (meters):
Mt. Everest, China (Tibet)–Nepal, 29,028 (8,848)
Lowest elevation in feet (meters):
Dead Sea, Israel–Jordan, 1,339 (408) below sea level

Australia

Area in square miles (sq km):
2,966,155 (7,682,300)
Estimated population:
19,455,000
Population per square mile (sq km):
6.6 (2.5)
Mean elevation in feet (meters):
1,000 (305)
Highest elevation in feet (meters):
Mt. Kosciuszko, New South Wales, 7,313 (2,229)
Lake Eyre, South Australia, 52 (16) below sea level

Europe

Area in square miles (sq km):
3,800,000 (9,900,000)
Estimated population:
728,975,000
Population per square mile (sq km):
192 (74)
Mean elevation in feet (meters):
980 (300)
Highest elevation in feet (meters):
Gora El'brus, Russia, 18,510 (5,642)
Lowest elevation in feet (meters):
Caspian Sea, Asia-Europe, 92 (28) below sea level

North America

Area in square miles (sq km):
9,500,000 (24,700,000)
Estimated population:
488,780,000
Population per square mile (sq km):
51 (20)
Mean elevation in feet (meters):
2,000 (610)
Highest elevation in feet (meters):
Mt. McKinley, Alaska, U.S., 20,320 (6,194)
Lowest elevation in feet (meters):
Death Valley, California, U.S., 282 (86) below sea level

Oceania (incl. Australia)

Area in square miles (sq km):
3,300,000 (8,500,000)
Estimated population:
31,415,000
Population per square mile (sq km):
9.5 (3.7)
Mean elevation in feet (meters):
0 (0)
Highest elevation in feet (meters):
Mt. Wilhelm, Papua New Guinea, 14,793 (4,509)
Lowest elevation in feet (meters):
Lake Eyre, South Australia, 52 (16) below sea level

South America

Area in square miles (sq km):
6,900,000 (17,800,000)
Estimated population:
352,960,000
Population per square mile (sq km):
51 (20)
Mean elevation in feet (meters):
1,800 (550)
Highest elevation in feet (meters):
Cerro Aconcagua, Argentina, 22,831 (6,959)
Lowest elevation in feet (meters):
Salinas Chicas, Argentina, 138 (42) below sea level

World

Area in square miles (sq km):
57,900,000 (150,100,000)
Estimated population:
6,195,885,000
Population per square mile (sq km):
107 (41)
Mean elevation in feet (meters):
0 (0)
Highest elevation in feet (meters):
Mt. Everest, China (Tibet)–Nepal, 29,028 (8,848)
Lowest elevation in feet (meters):
Dead Sea, Israel–Jordan, 1,339 (408) below sea level

Largest Islands

Rank	Name	Area square miles	square km
1	Greenland, North America	840,000	2,175,600
2	New Guinea, Asia-Oceania	309,000	800,000
3	Borneo (Kalimantan), Asia	287,300	744,100
4	Madagascar, Africa	226,500	587,000
5	Baffin Island, Canada	195,928	507,451
6	Sumatra (Sumatera), Indonesia	182,860	473,606
7	Honshū, Japan	89,176	230,966
8	Great Britain, United Kingdom	88,795	229,978
9	Victoria Island, Canada	83,897	217,291
10	Ellesmere Island, Canada	75,767	196,236
11	Celebes (Sulawesi), Indonesia	73,057	189,216
12	South Island, New Zealand	57,708	149,463
13	Java (Jawa), Indonesia	51,038	132,187
14	North Island, New Zealand	44,332	114,821
15	Cuba, North America	42,800	110,800
16	Newfoundland, Canada	42,031	108,860
17	Luzon, Philippines	40,420	104,688
18	Iceland, Europe	39,800	103,000
19	Mindanao, Philippines	36,537	94,630
20	Ireland, Europe	32,600	84,400
21	Hokkaidō, Japan	32,245	83,515
22	Sakhalin, Russia	29,500	76,400
23	Hispaniola, North America	29,400	76,200
24	Banks Island, Canada	27,038	70,028
25	Tasmania, Australia	26,200	67,800
26	Sri Lanka, Asia	24,900	64,600
27	Devon Island, Canada	21,331	55,247
28	Berkner Island, Antarctica	20,005	51,829
29	Alexander Island, Antarctica	19,165	49,652
30	Tierra del Fuego, South America	18,600	48,200
31	Novaya Zemlya, north island, Russia	18,436	47,764
32	Kyūshū, Japan	17,129	44,363
33	Melville Island, Canada	16,274	42,149
34	Southampton Island, Canada	15,913	41,214
35	Axel Heiberg, Canada	15,498	40,151
36	Spitsbergen, Norway	15,260	39,523
37	New Britain, Papua New Guinea	14,093	36,500
38	Taiwan, Asia	13,900	36,000
39	Hainan Dao, China	13,100	34,000
40	Prince of Wales Island, Canada	12,872	33,339
41	Novaya Zemlya, south island, Russia	12,633	32,730
42	Vancouver Island, Canada	12,079	31,285
43	Sicily, Italy	9,926	25,709
44	Somerset Island, Canada	9,570	24,786
45	Sardinia, Italy	9,301	24,090
46	Bathurst Island, Canada	7,600	19,684
47	Shikoku, Japan	7,258	18,799
48	Ceram (Seram), Indonesia	7,191	18,625
49	North East Land, Norway	6,350	16,446
50	New Caledonia, Oceania	6,252	16,192
51	Prince Patrick Island, Canada	5,986	15,509
52	Timor, Asia	5,743	14,874
53	Sumbawa, Indonesia	5,549	14,377
54	Ostrov Oktyabr'skoy Revolyutsii, Russia	5,511	14,279
55	Flores, Indonesia	5,502	14,250
56	Samar, Philippines	5,100	13,080
57	King William Island, Canada	4,961	12,853
58	Negros, Philippines	4,907	12,710
59	Thurston Island, Antarctica	4,854	12,576
60	Palawan, Philippines	4,550	11,785

Islands, Islands, Everywhere

Four islands—Hokkaidō, Honshū, Kyūshū, and Shikoku—constitute 98% of Japan's total land area, but the country is actually comprised of more than 3,000 islands. Similarly, two islands—Great Britain and Ireland—make up 93% of the total land area of the British Isles, but the island group also includes more than 5,000 smaller islands.

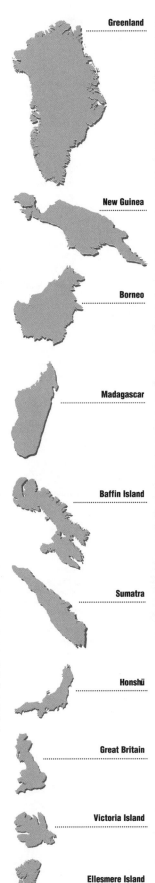

Major World Island Groups

Aleutian Islands (Pacific Ocean)
Alexander Archipelago (Pacific Ocean)
Azores (Atlantic Ocean)
Bahamas (Atlantic Ocean)
Balearic Islands (Mediterranean Sea)
Bismarck Archipelago (Pacific Ocean)
British Isles (Atlantic Ocean)
Cape Verde Islands (Atlantic Ocean)
Dodecanese (Mediterranean Sea)
Faroe Islands (Atlantic Ocean)
Falkland Islands (Atlantic Ocean)
Fiji Islands (Pacific Ocean)
Galapagos Islands (Pacific Ocean)
Greater Sunda Islands (Indian/Pacific Oceans)
Hawaiʻian Islands (Pacific Ocean)
Ionian Islands (Mediterranean Sea)
Islas Canarias (Atlantic Ocean)
Japan (Pacific Ocean)
Kikládhes (Mediterranean Sea)
Kuril Islands (Pacific Ocean)
Lesser Sunda Islands (Indian Ocean)
Moluccas (Pacific Ocean)
Nansei Shotō (Pacific Ocean)
New Hebrides (Atlantic Ocean)
New Siberian Islands (Arctic Ocean)
Novaya Zemlya (Arctic Ocean)
Philippine Islands (Pacific Ocean)
Severnaya Zemlya (Arctic Ocean)
Solomon Islands (Pacific Ocean)
Spitsbergen (Arctic Ocean)
West Indies (Atlantic Ocean)

Contrasting Population Densities

Some islands are among the most densely populated places on Earth, while others are among the least densely populated. This fact is dramatically illustrated by the following comparison of five islands:

Manhattan, N.Y., U.S., (pop. 1,537,000) — 69,864/ sq mile (26,965/ sq km)
Singapore Island, Singapore (pop. 4,375,000) — 17,785/ sq mile (6,879/ sq km)
Long Island, N.Y., U.S. (pop. 7,449,000) — 5,410/ sq mile (2,089/ sq km)

Population per square mile (sq km)

Baffin Island, Canada (pop. 11,700) 0.06/ sq mile (0.02/ sq km)
Greenland, (pop. 56,000) 0.07/ sq mile (0.03/ sq km)

Mountains, Volcanoes, and Earthquakes

The Tallest Mountain in the World

With its peak reaching 29,028 feet (8,848 m) above sea level, Mt. Everest ranks as the *highest* mountain in the world, but not the *tallest*. That title goes to Mauna Kea, one of the five volcanic mountains that make up the island of Hawai'i. From its base on the floor of the Pacific Ocean, Mauna Kea rises 33,476 feet (10,210 m)—more than six miles—although only the top 13,796 feet (4,205 m) are above sea level.

Seafloor Atop Mt. Everest

When Sir Edmund Percival Hillary and Tenzing Norgay reached the summit of Mt. Everest in 1953, they probably did not realize they were standing on the seafloor.

The Himalayan mountain system was formed through the process of plate tectonics. Ocean once separated India and Asia, but 180 million years ago the Indo-Australian crustal plate, on which India sits, began a northward migration and eventually collided with the Eurasian plate. The seafloor between the two landmasses crumpled and was slowly thrust upward. Rock layers that once lay at the bottom of the ocean now crown the peaks of the highest mountains in the world.

Principal Mountain Systems and Ranges of the World

Alaska Range (North America)
Alps (Europe)
Altai (Asia)
Andes (South America)
Appennino (Europe)
Atlas Mountains (Africa)
Appalachian Mountains (North America)
Brooks Range (North America)
Carpathian Mountains (Europe)
Cascade Range (North America)
Caucasus (Europe/Asia)
Coast Mountains (North America)
Coast Ranges (North America)
Great Dividing Range (Australia)
Greater Khingan Range (Asia)
Himalayas (Asia)
Hindu Kush (Asia)
Karakoram Range (Asia)
Kunlun Shan (Asia)
Madre Occidental, Sierra (North America)
Madre Oriental, Sierra (North America)
Nevada, Sierra (North America)
Pamirs (Asia)
Pyrenees (Europe)
Rocky Mountains (North America)
Sayan Khrebet (Asia)
Southern Alps (New Zealand)
Tien Shan (Asia)
Urals (Europe)
Zagros Mountains (Asia)

Principal Mountains of the World Δ = *Highest mountain in range, region, country, or state named*

Location	Height Feet	Meters
Africa		
Kilimanjaro, Δ Tanzania (Δ Africa)	19,340	5,895
Kirinyaga (Mount Kenya), Δ Kenya	17,058	5,199
Margherita Peak, Δ Uganda-Δ Dem. Rep. of the Congo	16,763	5,109
Ras Dashen Terara, Δ Ethiopia	15,158	4,620
Meru, Mount, Tanzania	14,978	4,565
Karisimbi, Volcan, Δ Rwanda-Dem. Rep. of the Congo	14,787	4,507
Elgon, Mount, Kenya-Uganda	14,178	4,321
Toubkal, Jebel, Δ Morocco (Δ Atlas Mts.)	13,665	4,165
Cameroon Mountain, Δ Cameroon	13,451	4,100
Antarctica		
Vinson Massif, Δ Antarctica	16,066	4,897
Kirkpatrick, Mount	14,856	4,528
Markham, Mount	14,049	4,282
Jackson, Mount	13,747	4,190
Sidley, Mount	13,717	4,181
Wade, Mount	13,396	4,083
Asia		
Everest, Mount, Δ China-Δ Nepal (Δ Tibet; Δ Himalayas; Δ Asia; Δ World)	29,028	8,848
K2 (Qogir Feng), China-Δ Pakistan (Δ Kashmir; Δ Karakoram Range)	28,250	8,611
Kanchenjunga, Δ India-Nepal	28,208	8,598
Makalu, China-Nepal	27,825	8,481
Dhawalāgiri, Nepal	26,810	8,172
Nanga Parbat, Pakistan	26,660	8,126
Annapurna, Nepal	26,504	8,078
Gasherbrum, China-Pakistan	26,470	8,068
Xixabangma Feng, China	26,286	8,012
Nanda Devi, India	25,645	7,817
Kamet, China-India	25,447	7,756
Namjagbarwa Feng, China	25,446	7,756
Muztag, China (Δ Kunlun Shan)	25,338	7,723
Tirich Mir, Pakistan (Δ Hindu Kush)	25,230	7,690
Gongga Shan, China	24,790	7,556
Kula Kangri, Δ Bhutan	24,784	7,554
Ismail Samani, pik, Δ Tajikistan (Δ Pamir)	24,590	7,495
Nowshak, Δ Afghanistan-Pakistan	24,557	7,485
Pobedy, Pik, China-Russia	24,406	7,439
Chomo Lhari, Bhutan-China	23,997	7,314
Muztag, China	23,891	7,282
Lenina, Pik, Δ Kyrgyzstan-Tajikistan	23,406	7,134
Api, Nepal	23,399	7,132
Kangrinboqê Feng, China	22,028	6,714
Hkakabo Razi, Δ Myanmar	19,296	5,881
Damavand, Qolleh-ye, Δ Iran	18,386	5,604
Agri Dagi (Mount Ararat), Δ Turkey	16,854	5,137
Fuladi, Kuh-e, Afghanistan	16,847	5,135
Jaya, Puncak, Δ Indonesia (Δ New Guinea)	16,503	5,030
Klyuchevskaya, Vulkan, Russia (Δ Poluostrov Kamchatka)	15,584	4,750
Trikora, Puncak, Indonesia	15,584	4,750
Belukha, Gora, Kazakhstan-Russia	14,783	4,506
Turgen, Mount, Mongolia	14,311	4,362
Kinabalu, Gunong, Δ Malaysia (Δ Borneo)	13,455	4,101
Yü Shan, Δ Taiwan	13,114	3,997
Erciyes Dagı, Turkey	12,851	3,917
Kerinci, Gunung, Indonesia (Δ Sumatra)	12,467	3,800
Fuji San, Δ Japan (Δ Honshu)	12,388	3,776
Rinjani, Gunung, Indonesia (Δ Lombok)	12,224	3,726
Semeru, Gunung, Indonesia (Δ Java)	12,060	3,676
Hadūr Shu'ayb, Jabal an-, Δ Yemen (Δ Arabian Peninsula)	12,008	3,660
Australia / Oceania		
Wilhelm, Mt., Δ Papua New Guinea	14,793	4,509
Giluwe, Mt., Papua New Guinea	14,330	4,368
Bangeta, Mt., Papua New Guinea	13,520	4,121
Victoria, Mt., Papua New Guinea (Δ Owen Stanley Range)	13,238	4,035
Aoraki (Mt. Cook), Δ New Zealand (Δ South Island)	12,316	3,754
Europe		
El'brus, Gora, Δ Russia (Δ Caucasus; Δ Europe)	18,510	5,642
Dykhtau, Mt., Russia	17,073	5,204
Blanc, Mont (Monte Bianco) Δ France-Δ Italy (Δ Alps)	15,771	4,807

Location	Height Feet	Meters
Dufourspitze, Italy-Δ Switzerland	15,203	4,634
Weisshorn, Switzerland	14,783	4,506
Matterhorn, Italy-Switzerland	14,692	4,478
Finsteraarhorn, Switzerland	14,022	4,274
Jungfrau, Switzerland	13,642	4,158
Écrins, Barre des, France	13,458	4,102
Viso, Monte, Italy (Δ Cottian Alps)	12,602	3,841
Grossglockner, Δ Austria	12,461	3,798
Teide, Pico de, Δ Spain (Δ Canary Is.)	12,188	3,715
North America		
McKinley, Mt., Δ Alaska (Δ United States; Δ North America)	20,320	6,194
Logan, Mt., Δ Canada (Δ Yukon; Δ St. Elias Mts.)	19,551	5,959
Orizaba, Pico de, Δ Mexico	18,406	5,610
St. Elias, Mt., Alaska-Canada	18,008	5,489
Popocatépetl, Volcán, Mexico	17,930	5,465
Foraker, Mt., Alaska	17,400	5,304
Iztaccíhuatl, Mexico	17,159	5,230
Lucania, Mt., Canada	17,147	5,226
Fairweather, Mt., Alaska-Canada (Δ British Columbia)	15,300	4,663
Whitney, Mt., Δ California	14,494	4,418
Elbert, Mt., Δ Colorado (Δ Rocky Mts.)	14,433	4,399
Massive, Mt., Colorado	14,421	4,396
Harvard, Mt., Colorado	14,420	4,395
Rainier, Mt., Δ Washington (Δ Cascade Range)	14,410	4,392
Williamson, Mt., California	14,370	4,380
La Plata Pk., Colorado	14,361	4,377
Blanca Pk., Colorado (Δ Sangre de Cristo Mts.)	14,345	4,372
Uncompahgre Pk., Colorado (Δ San Juan Mts.)	14,309	4,361
Grays Pk., Colorado (Δ Front Range)	14,270	4,349
Evans, Mt., Colorado	14,264	4,348
Longs Pk., Colorado	14,255	4,345
Wrangell, Mt., Alaska	14,163	4,317
Shasta, Mt., California	14,162	4,317
Pikes Pk., Colorado	14,110	4,301
Colima, Nevado de, Mexico	13,991	4,240
Tajumulco, Volcán, Δ Guatemala (Δ Central America)	13,845	4,220
Gannett Pk., Δ Wyoming	13,804	4,207
Mauna Kea, Δ Hawaii	13,796	4,205
Grand Teton, Wyoming	13,770	4,197
Mauna Loa, Hawaii	13,679	4,169
Kings Pk., Δ Utah	13,528	4,123
Cloud Pk., Wyoming (Δ Bighorn Mts.)	13,167	4,013
Waddington, Mt., Canada (Δ Coast Mts.)	13,163	4,012
Wheeler Pk., Δ New Mexico	13,161	4,011
Boundary Pk., Δ Nevada	13,140	4,005
Robson, Mt., Canada (Δ Canadian Rockies)	12,972	3,954
Granite Pk., Δ Montana	12,799	3,901
Borah Pk., Δ Idaho	12,662	3,859
Humphreys Pk., Δ Arizona	12,633	3,851
Chirripó, Volcán, Δ Costa Rica	12,530	3,819
Columbia, Mt., Canada (Δ Alberta)	12,294	3,747
Adams, Mt., Washington	12,276	3,742
Gunnbjørn Fjeld, Δ Greenland	12,139	3,700
South America		
Aconcagua, Cerro, Δ Argentina (Δ Andes; Δ South America)	22,831	6,959
Ojos del Salado, Nevado, Argentina-Δ Chile	22,615	6,893
Bonete, Cerro, Argentina	22,546	6,872
Huascarán, Nevado, Δ Peru	22,133	6,746
Llullaillaco, Volcán, Argentina-Chile	22,110	6,739
Yerupaja, Nevado, Peru	21,765	6,634
Tupungato, Cerro, Argentina-Chile	21,555	6,570
Sajama, Nevado, Bolivia	21,463	6,542
Illampu, Nevado, Bolivia	21,066	6,421
Illimani, Nevado, Bolivia	20,741	6,322
Chimborazo, Δ Ecuador	20,702	6,310
Antofalla, Volcán, Argentina	20,013	6,100
Cotopaxi, Ecuador	19,347	5,897
Misti, Volcán, Peru	19,101	5,822
Huila, Nevado de, Colombia (Δ Cordillera Central)	18,865	5,750
Bolívar, Pico, Δ Venezuela	16,427	5,007

Notable Volcanic Eruptions

Year	Volcano Name, Location	Comments
ca. 4895 B.C.	Crater Lake, Oregon, U.S.	Collapse forms caldera that now contains Crater Lake.
ca. 4350 B.C.	Kikai, Ryukyu Islands, Japan	Japan's largest known eruption.
ca. 1628 B.C.	Santorini (Thira), Greece	Eruption devastates late Minoan civilization.
79 A.D.	Vesuvius (Vesuvio), Italy	Roman towns of Pompeii and Herculaneum are buried.
ca. 180	Taupo, New Zealand	Area measuring 6,200 square miles (16,000 sq km) is devastated.
ca. 260	Ilopango, El Salvador	Thousands killed, with major impact on Mayan civilization.
915	Towada, Honshu, Japan	Japan's largest historic eruption.
ca. 1000	Baitoushan, China/Korea	Largest known eruption on Asian mainland.
1259	Unknown	Evidence from polar ice cores suggests that a huge eruption, possibly the largest of the millennium, occurred in this year.
1586	Kelut, Java	Explosions in crater lake; mudflows kill 10,000.
1631	Vesuvius (Vesuvio), Italy	Eruption kills 4,000.
ca. 1660	Long Island, Papua New Guinea	"The time of darkness" in tribal legends on Papua New Guinea.
1672	Merapi, Java	Pyroclastic flows and mudflows kill 3,000.
1711	Awu, Sangihe Islands, Indonesia	Pyroclastic flows kill 3,000.
1760	Makian, Halmahera, Indonesia	Eruption kills 2,000; island evacuated for seven years.
1772	Papandayan, Java	Debris avalanche causes 2,957 fatalities.
1783	Lakagigar, Iceland	Largest historic lava flows; 9,350 deaths.
1790	Kīlauea, Hawai'i	Hawai'i's last large explosive eruption.
1792	Unzen, Kyushu, Japan	Tsunami and debris avalanche kill 14,500.
1815	Tambora, Indonesia	History's most explosive eruption; 92,000 deaths.
1822	Galunggung, Java	Pyroclastic flows and mudflows kill 4,011.
1856	Awu, Sangihe Islands, Indonesia	Pyroclastic flows kill 2,806.
1883	Krakatau, Indonesia	Caldera collapse; 36,417 people killed, most by tsunami.
1888	Ritter Island, Papua New Guinea	3,000 killed, most by tsunami created by debris avalanche.
1902	Mont Pelee, West Indies	Town of St. Pierre destroyed; 28,000 people killed.
1902	Santa Maria, Guatemala	5,000 killed as 10 villages are buried by volcanic debris.
1912	Novarupta (Katmai), Alaska	Largest 20th-century eruption.
1914	Lassen, California, U.S.	California's last historic eruption.
1919	Kelut, Java	Mudflows devastate 104 villages and kill 5,110 people.
1930	Merapi, Java	1,369 people are killed as 42 villages are totally or partially destroyed.
1943	Parícutin, Mexico	Fissure in cornfield erupts, building cinder cone 1,500 feet (460 m) high within two years. One of the few volcano births ever witnessed.
1951	Lamington, Papua New Guinea	Pyroclastic flows kill 2,942.
1963	Surtsey, Iceland	Submarine eruption builds new island.
1977	Nyiragongo, Dem. Rep. of the Congo	One of the shortest major eruptions and fastest lava flows ever recorded.
1980	St. Helens, Washington, U.S.	Lateral blast; 230-square-mile (600 sq km) area devastated.
1982	El Chichón, Mexico	Pyroclastic surges kill 1,877.
1985	Ruiz, Colombia	Mudflows kill 23,080.
1991	Pinatubo, Luzon, Philippines	Major eruption in densely populated area prompts evacuation of 250,000 people; fatalities number fewer than 800. Enormous amount of gas released into stratosphere lowers global temperatures for more than a year.
1995	Soufriere Hills Volcano, Montserrat Island	Forced evacuation of the southern half of the island, destroyed capital city of Plymouth.

Sources: Smithsonian Institution Global Volcanism Program; Volcanoes of the World, Second Edition, by Tom Simkin and Lee Siebert, Geoscience Press and Smithsonian Institution, 1994. USGS National Earthquake Information Center.

Eruption of Mt. St. Helens in 1980

Significant Earthquakes through History

Year	Estimated Magnitude	Number of Deaths	Place
365		50,000	Knossos, Crete
844		50,000	Damascus, Syria; Antioch, Turkey
856		150,000	Dāmghān, Kashan, Qumis, Iran
893		150,000	Caucasus region
894		180,000	western India
1042		50,000	Palmyra, Baalbek, Syria
1138		230,000	Aleppo, Gansana, Syria
1139	6.8	300,000	Gāncä, Kiapas, Azerbaijan
1201		50,000	upper Egypt to Syria
1290	6.7	100,000	eastern China
1556		820,000	Shanxi Province, China
1662		300,000	China
1667	6.9	80,000	Caucusus region, northern Iran
1668		50,000	Shandong Province, China
1693		93,000	Sicily, Italy
1727		77,000	Tabrīz, Iran
1731		100,000	Beijing, China
1739		50,000	China
1755		62,000	Morocco, Portugal, Spain
1780	6.7	100,000	Tabrīz, Iran
1868	7.7	70,000	Ecuador, Colombia
1908	7.5	83,000	Calabria, Messina, Italy
1920	8.5	200,000	Gansu and Shanxi provinces, China

Year	Estimated Magnitude	Number of Deaths	Place
1923	8.2	142,807	Tokyo, Yokohama, Japan
1927	8.3	200,000	Gansu and Qinghai provinces, China
1932	7.6	70,000	Gansu Province, China
1970	7.8	66,794	northern Peru
1976	7.8	242,000	Tangshan, China
1990	7.7	50,000	northwestern Iran

Some Significant U.S. Earthquakes

Year	Estimated Magnitude	Number of Deaths	Place
1811–12	8.6, 8.4, 8.7	<10	New Madrid, Missouri (series)
1886	7.0	60	Charleston, South Carolina
1906	8.3	3,000	San Francisco, California
1933	6.3	115	Long Beach, California
1946	7.4	5 ‡	Alaska
1964	8.4	125	Anchorage, Alaska
1971	6.8	65	San Fernando, California
1989	7.1	62	San Francisco Bay Area, California
1994	6.8	58	Northridge, California

‡ *A tsunami generated by this earthquake struck Hilo, Hawaii, killing 159 people.*

Oceans and Lakes

Oceans, Seas, Gulfs, and Bays

	Area sq. miles	sq. km.	Volume of water cubic miles	cubic km.	Mean depth feet	meters	Greatest known depth feet	meters	
Pacific Ocean	63,800,000	165,200,000	169,650,000	707,100,000	12,987	3,957	35,810	10,922	Mariana Trench
Atlantic Ocean	31,800,000	82,400,000	79,199,000	330,100,000	11,821	3,602	28,232	8,611	Puerto Rico Trench
Indian Ocean	28,900,000	74,900,000	68,282,000	284,600,000	12,261	3,736	23,812	7,258	Weber Basin
Arctic Ocean	5,400,000	14,000,000	4,007,000	16,700,000	3,712	1,131	17,897	5,453	Lat. 77° 45'N, long. 175°W
Coral Sea	1,850,000	4,791,000	2,752,000	11,470,000	7,857	2,394	30,079	9,165	
Arabian Sea	1,492,000	3,864,000	2,416,000	10,070,000	8,973	2,734	19,029	5,803	
South China Sea	1,331,000	3,447,000	943,000	3,929,000	3,741	1,140	18,241	5,563	
Caribbean Sea	1,063,000	2,753,000	1,646,000	6,860,000	8,175	2,491	25,197	7,685	Off Cayman Islands
Mediterranean Sea	967,000	2,505,000	901,000	3,754,000	4,916	1,498	16,470	5,023	Off Cape Matapan, Greece
Bering Sea	876,000	2,269,000	911,000	3,796,000	5,382	1,640	25,194	7,684	Off Buldir Island
Bengal, Bay of	839,000	2,173,000	1,357,000	5,616,000	8,484	2,585	17,251	5,261	
Okhotsk, Sea of	619,000	1,603,000	316,000	1,317,000	2,694	821	1,029	3,374	Lat. 146° 10'E, long. 46° 50'N
Norwegian Sea	597,000	1,546,000	578,000	2,408,000	5,717	1,742	13,189	4,022	
Mexico, Gulf of	596,000	1,544,000	560,000	2,332,000	8,205	2,500	14,370	4,382	Sigsbee Deep
Hudson Bay	475,000	1,230,000	22,000	92,000	328	100	850	259	Near entrance
Greenland Sea	465,000	1,204,000	417,000	1,740,000	4,739	1,444	15,899	4,849	
Japan, Sea of	413,000	1,070,000	391,000	1,630,000	5,037	1,535	12,041	3,669	
Arafura Sea	400,000	1,037,000	49,000	204,000	646	197	12,077	3,680	
East Siberian Sea	357,000	926,000	14,000	61,000	216	66	508	155	
Kara Sea	349,000	903,000	24,000	101,000	371	113	2,034	620	
East China Sea	290,000	752,000	63,000	263,000	1,145	349	7,778	2,370	
Banda Sea	268,000	695,000	511,000	2,129,000	10,056	3,064	24,418	7,440	
Baffin Bay	263,000	681,000	142,000	593,000	2,825	861	7,010	2,136	
Laptev Sea	262,000	678,000	87,000	363,000	1,772	540	9,780	2,980	
Timor Sea	237,000	615,000	60,000	250,000	1,332	406	10,863	3,310	
Andaman Sea	232,000	602,000	158,000	660,000	3,597	1,096	13,777	4,198	
Chukchi Sea	228,000	590,000	11,000	45,000	252	77	525	160	
North Sea	214,000	554,000	12,000	52,000	315	96	2,655	809	
Java Sea	185,000	480,000	5,000	22,000	147	45	292	89	
Beaufort Sea	184,000	476,000	115,000	478,000	3,295	1,004	12,245	3,731	
Red Sea	174,000	450,000	60,000	251,000	1,831	558	8,648	2,635	
Baltic Sea	173,000	448,000	5,000	20,000	157	48	1,506	459	
Celebes Sea	168,000	435,000	380,000	1,586,000	11,962	3,645	19,173	5,842	
Black Sea	166,000	431,000	133,000	555,000	3,839	1,170	7,256	2,211	
Yellow Sea	161,000	417,000	4,000	17,000	131	40	344	105	
Sulu Sea	134,000	348,000	133,000	553,000	5,221	1,591	18,300	5,576	
Molucca Sea	112,000	291,000	133,000	554,000	6,242	1,902	16,311	4,970	
Ceram Sea	72,000	187,000	54,000	227,000	3,968	1,209	17,456	5,319	
Flores Sea	47,000	121,000	53,000	222,000	6,003	1,829	16,813	5,123	
Bali Sea	46,000	119,000	12,000	49,000	1,349	411	4,253	1,296	
Savu Sea	41,000	105,000	43,000	178,000	5,582	1,701	11,060	3,370	
White Sea	35,000	91,000	1,000	4,400	161	49	1,083	330	
Azov, Sea of	15,000	40,000	100	400	29	9	46	14	
Marmara, Sea of	4,000	11,000	1,000	4,000	1,171	357	4,138	1,261	

Source: Atlas of World Water Balance, *USSR National Committee for the International Water Decade and UNESCO, 1977.*

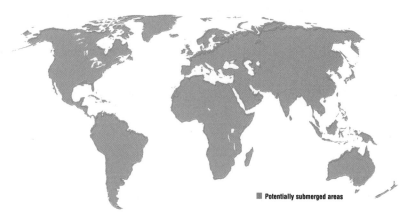

■ Potentially submerged areas

Fluctuating Sea Level

Changes in the Earth's climate have a dramatic effect on the sea level. Only 20,000 years ago, at the height of the most recent ice age, a vast amount of the Earth's water was locked up in ice sheets and glaciers, and the sea level was 330 feet (100 meters) lower than it is today. As the climate warmed slowly, the ice began to melt and the oceans began to rise.

Today there is still a tremendous amount of ice on the Earth. More than nine-tenths of it resides in the enormous ice cap which covers Antarctica. Measuring about 5.4 million square miles (14 million sq km) in surface area, the ice cap is on average one mile (1.6 km) thick but in some places is nearly three miles (4.8 km) thick. If it were to melt, the oceans would rise another 200 feet (60 m), and more than half of the world's population would have to relocate.

Ocean Depths in Profile

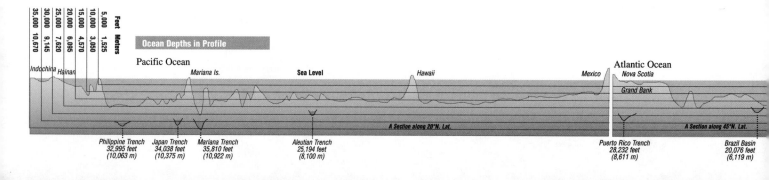

Deepest Lakes

	Lake	Greatest depth feet	meters
1	Baikal, Lake, Russia	5,315	1,621
2	Tanganyika, Lake, Africa	4,800	1,464
3	Caspian Sea, Asia-Europe	3,363	1,025
4	Nyasa, Lake (Lake Malawi), Malawi-Mozambique-Tanzania	2,317	706
5	Issyk-Kul', Lake, Kyrgyzstan	2,303	702
6	Great Slave Lake, NWT, Canada	2,015	614
7	Matana, Lake, Indonesia	1,936	590
8	Crater Lake, Oregon, U.S.	1,932	589
9	Toba, Lake (Danau Toba), Indonesia	1,736	529
10	Sarez, Lake, Tajikistan	1,657	505
11	Tahoe, Lake, California-Nevada, U.S.	1,645	502
12	Kivu, Lake, Rwanda-Dem. Rep. of the Congo	1,628	496
13	Chelan, Lake, Washington, U.S.	1,605	489
14	Quesnel Lake, BC, Canada	1,560	476
15	Adams Lake, BC, Canada	1,500	457

Lakes with the Greatest Volume of Water

	Lake	Volume of water cubic mi	cubic km
1	Caspian Sea, Asia-Europe	18,900	78,200
2	Baikal, Lake, Russia	5,500	23,000
3	Tanganyika, Lake, Africa	4,500	18,900
4	Superior, Lake, Canada-U.S.	2,900	12,200
5	Nyasa, Lake (Lake Malawi), Malawi-Mozambique-Tanzania	1,900	7,725
6	Michigan, Lake, U.S.	1,200	4,910
7	Huron, Lake, Canada-U.S.	860	3,580
8	Victoria, Lake, Kenya-Tanzania-Uganda	650	2,700
9	Issyk-Kul', Lake, Kyrgyzstan	415	1,730
10	Ontario, Lake, Canada-U.S.	410	1,710
11	Great Slave Lake, Canada	260	1,070
12	Great Bear Lake, Canada	240	1,010
13	Ladozhskoye, Ozero, Russia	220	908
14	Titicaca, Lago, Bolivia-Peru	170	710

Sources for volume and depth information: Atlas of World Water Balance, *USSR National Committee for the International Water Decade and UNESCO, 1977;* Principal Rivers and Lakes of the World, *National Oceanic and Atmospheric Administration, 1982.*

Principal Lakes

	Lake	Area sq mi	sq km
1	Caspian Sea, Asia-Europe	143,240	370,990
2	Superior, Lake, Canada-U.S.	31,700	82,100
3	Victoria, Lake, Kenya-Tanzania-Uganda	26,820	69,463
4	Huron, Lake, Canada-U.S.	23,000	60,000
5	Michigan, Lake, U.S.	22,300	57,800
6	Tanganyika, Lake, Africa	12,350	31,986
7	Baikal, Lake, Russia	12,200	31,500
8	Great Bear Lake, Canada	12,095	31,326
9	Nyasa, Lake (Lake Malawi), Malawi-Mozambique-Tanzania	11,150	28,878
10	Aral Sea, Kazakhstan-Uzbekistan	11,100	28,700
11	Great Slave Lake, Canada	11,030	28,568
12	Erie, Lake, Canada-U.S.	9,910	25,667
13	Winnipeg, Lake, Canada	9,416	24,387
14	Ontario, Lake, Canada-U.S.	7,540	19,529
15	Balqash koli (Lake Balkhash), Kazakhstan	7,100	18,300
16	Ladozhskoye, Ozero, Russia	6,833	17,700
17	Chad, Lake (Lac Tchad), Cameroon-Chad-Nigeria	6,300	16,300
18	Onezhskoye, Ozero, Russia	3,753	9,720
19	Eyre, Lake, Australia	3,700	9,500
20	Titicaca, Lago, Bolivia-Peru	3,200	8,300
21	Nicaragua, Lago de, Nicaragua	3,150	8,158
22	Mai-Ndombe, Lac, Dem. Rep. of the Congo	3,100	8,000
23	Athabasca, Lake, Canada	3,064	7,935
24	Reindeer Lake, Canada	2,568	6,650
25	Tônlé Sap, Cambodia	2,500	6,500
26	Rudolf, Lake, Ethiopia-Kenya	2,473	6,405
27	Issyk-Kul', Ozero, Kyrgyzstan	2,425	6,280
28	Torrens, Lake, Australia	2,300	5,900
29	Albert, Lake, Uganda-Dem. Rep. of the Congo	2,160	5,594
30	Vänern, Sweden	2,156	5,584
31	Nettilling Lake, Canada	2,140	5,542
32	Winnipegosis, Lake, Canada	2,075	5,374
33	Bangweulu, Lake, Zambia	1,930	4,999
34	Nipigon, Lake, Canada	1,872	4,848
35	Orumiyeh, Daryacheh-ye, Iran	1,815	4,701
36	Manitoba, Lake, Canada	1,785	4,624
37	Woods, Lake of the, Canada-U.S.	1,727	4,472
38	Kyoga, Lake, Uganda	1,710	4,429

Lake Baikal

Russia's Great Lake

On a map of the world, Lake Baikal is easy to overlook — a thin blue crescent adrift in the vastness of Siberia. But its inconspicuousness is deceptive, for Baikal is one of the greatest bodies of fresh water on Earth.

Although lakes generally have a life span of less than one million years, Baikal has existed for perhaps as long as 25 million years, which makes it the world's oldest body of fresh water. It formed in a rift that tectonic forces had begun to tear open in the Earth's crust. As the rift grew, so did Baikal. Today the lake is 395 miles (636 km) long and an average of 30 miles (48 km) wide. Only seven lakes in the world have a greater surface area.

Baikal is the world's deepest lake. Its maximum depth is 5,315 feet (1,621 m) — slightly over a mile, and roughly equal to the greatest depth of the Grand Canyon. The lake bottom lies 4,250 feet (1,295 m) below sea level and two-and-a-third miles (3.75 km) below the peaks of the surounding mountains. The crustal rift which Baikal occupies is the planet's deepest land depression, extending to a depth of more than five-and-a-half miles (9 km). The lake sits atop at least four miles (6.4 km) of sediment, the accumulation of 25 million years.

More than 300 rivers empty into Baikal, but only one, the Angara, flows out of it. Despite having only 38% of the surface area of North America's Lake Superior, Baikal contains more water than all five of the Great Lakes combined. Its volume of 5,500 cubic miles (23,000 cubic km) is greater than that of any other freshwater lake in the world and represents approximately one-fifth of all of the Earth's unfrozen fresh water.

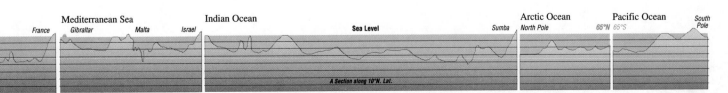

Rivers

World's Longest Rivers

Rank	River	Length (Miles)	Length (Kilometers)	Rank	River	Length (Miles)	Length (Kilometers)
1	Nile, Africa	4,145	6,671	36	Murray, Australia	1,566	2,520
2	Amazon (Amazonas)-Ucayali, South America	4,000	6,400	37	Ganges, Asia	1,560	2,511
3	Yangtze (Chang), Asia	3,900	6,300	38	Pilcomayo, South America	1,550	2,494
4	Mississippi-Missouri, North America	3,740	6,019	39	Euphrates, Asia	1,510	2,430
5	Huang (Yellow), Asia	3,395	5,464	40	Ural, Asia	1,509	2,428
6	Ob'-Irtysh, Asia	3,362	5,410	41	Arkansas, North America	1,459	2,348
7	Río de la Plata-Paraná, South America	3,030	4,876	42	Colorado, North America (U.S.-Mexico)	1,450	2,334
8	Congo, Africa	2,900	4,700	43	Aldan, Asia	1,412	2,273
9	Paraná, South America	2,800	4,500	44	Syr Darya, Asia	1,370	2,205
10	Amur-Argun, Asia	2,761	4,444	45	Dnieper, Europe	1,350	2,200
11	Lena, Asia	2,700	4,400	46	Araguaia, South America	1,350	2,200
12	Mackenzie, North America	2,635	4,241	47	Cassai (Kasai), Africa	1,338	2,153
13	Mekong, Asia	2,600	4,200	48	Tarim, Asia	1,328	2,137
14	Niger, Africa	2,600	4,200	49	Kolyma, Asia	1,323	2,129
15	Yenisey, Asia	2,543	4,092	50	Orange, Africa	1,300	2,100
16	Missouri-Red Rock, North America	2,533	4,076	51	Negro, South America	1,300	2,100
17	Mississippi, North America	2,348	3,779	52	Ayeyarwady (Irrawaddy), Asia	1,300	2,100
18	Murray-Darling, Australia	2,330	3,750	53	Red, North America	1,270	2,044
19	Missouri, North America	2,315	3,726	54	Juruá, South America	1,250	2,012
20	Volga, Europe	2,194	3,531	55	Columbia, North America	1,240	2,000
21	Madeira, South America	2,013	3,240	56	Xingu, South America	1,230	1,979
22	São Francisco, South America	1,988	3,199	57	Ucayali, South America	1,220	1,963
23	Grande, Rio (Río Bravo), North America	1,885	3,034	58	Saskatchewan-Bow, North America	1,205	1,939
24	Purús, South America	1,860	2,993	59	Peace, North America	1,195	1,923
25	Indus, Asia	1,800	2,900	60	Tigris, Asia	1,180	1,899
26	Danube, Europe	1,776	2,858	61	Don, Europe	1,162	1,870
27	Brahmaputra, Asia	1,770	2,849	62	Songhua, Asia	1,140	1,835
28	Yukon, North America	1,770	2,849	63	Pechora, Europe	1,124	1,809
29	Salween (Nu), Asia	1,750	2,816	64	Kama, Europe	1,122	1,805
30	Zambezi, Africa	1,700	2,700	65	Limpopo, Africa	1,120	1,800
31	Vilyuy, Asia	1,647	2,650	66	Angara, Asia	1,105	1,779
32	Tocantins, South America	1,640	2,639	67	Snake, North America	1,038	1,670
33	Orinoco, South America	1,615	2,600	68	Uruguay, South America	1,025	1,650
34	Paraguay, South America	1,610	2,591	69	Churchill, North America	1,000	1,600
35	Amu Darya, Asia	1,578	2,540	70	Marañón, South America	995	1,592

The World's Greatest River

Although the Nile is slightly longer, the Amazon surpasses all other rivers in volume, size of drainage basin, and in nearly every other important category. If any river is to be called the greatest in the world, surely it is the Amazon.

It has been estimated that one-fifth of all of the flowing water on Earth is carried by the Amazon. From its 150-mile (240-km)-wide mouth, the river discharges 6,180,000 cubic feet (174,900 cubic m) of water per second — four-and-a-half times as much as the Congo, ten times as much as the Mississippi, and fifty-six times as much as the Nile. The Amazon's tremendous outflow turns the waters of the Atlantic from salty to brackish for more than 100 miles (160 km) offshore.

Covering more than one-third of the entire continent of South America, the Amazon's vast drainage basin measures 2,669,000 square miles (6,915,000 sq km) and is nearly twice as large as that of the second-ranked Congo. The Amazon begins its 4,000-mile (6,400-km) journey to the Atlantic from high up in the Andes, only 100 miles (160 km) from the Pacific. Along its course it receives the waters of more than 1,000 tributaries, which rise principally from the Andes, the Guiana Highlands, and the Brazilian Highlands. Seven of the tributaries are more than 1,000 miles (1,600 km) long, and one, the Madeira, is more than 2,000 miles (3,200 km) long.

The depth of the Amazon throughout most of its Brazilian segment exceeds 150 feet (45 m). Depths of more than 300 feet (90 m) have been recorded at points near the mouth. The largest ocean-going vessels can sail as far inland as Manaus, 1,000 miles (1,600 km) from the mouth. Freighters and small passenger vessels can navigate to Iquitos, 2,300 miles (3,700 km) from the mouth, even during times of low water.

Drainage basin of the Amazon River

Rivers with the Greatest Volume of Water

Rank	River Name	Flow of water per second at mouth — cubic feet	cubic meters	Rank	River Name	Flow of water per second at mouth — cubic feet	cubic meters
1	Amazon (Amazonas), South America	6,180,000	174,900	18	Para-Tocantins, South America (joins Amazon at mouth)	360,000	10,200
2	Congo, Africa	1,377,000	39,000	19	Salween, Asia	353,000	10,000
3	Negro, South America (tributary of Amazon)	1,236,000	35,000	20	Cassai (Kasai), Africa (trib. of Congo)	351,000	9,900
4	Orinoco, South America	890,000	25,200	21	Mackenzie, North America	343,000	9,700
5	Río de la Plata–Paraná, South America	809,000	22,900	22	Volga, Europe	271,000	7,700
6	Yangtze (Chang), Asia;	770,000	21,800	23	Ohio, North America (trib. of Mississippi)	257,000	7,300
	Madeira, South America (trib. of Amazon)	770,000	21,800	24	Yukon, North America	240,000	6,800
7	Missouri, North America (trib. of Mississippi)	763,000	21,600	25	Indus, Asia	235,000	6,600
8	Mississippi, North America*	640,300	18,100	26	Danube, Europe	227,000	6,400
9	Yenisey, Asia	636,000	18,000	27	Niger, Africa	215,000	6,100
10	Brahmaputra, Asia	575,000	16,300	28	Atchafalaya, North America	181,000	5,100
11	Lena, Asia	569,000	16,100	29	Paraguay, South America	155,000	4,400
12	Zambezi, Africa	565,000	16,000	30	Ob'-Katun, Asia	147,000	4,200
13	Mekong, Asia	500,000	14,100	31	São Francisco, South America	120,000	3,400
14	Saint Lawrence, North America	460,000	13,000	32	Tunguska, Asia	118,000	3,350
15	Ayeyarwady (Irrawaddy), Asia	447,000	12,600	33	Huang (Yellow), Asia	116,000	3,300
16	Ob'-Irtysh, Asia; Ganges, Asia	441,000	12,500	34	Nile, Africa	110,000	3,100
17	Amur, Asia	390,000	11,000				

*Approximately one-third of the Mississippi's water is diverted above Baton Rouge, Louisiana, and reaches the Gulf of Mexico via the Atchafalaya River.

Principal Rivers of the Continents

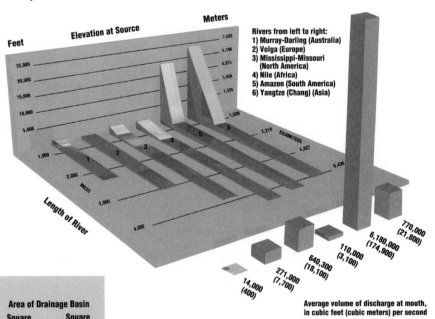

Rivers from left to right:
1) Murray-Darling (Australia)
2) Volga (Europe)
3) Mississippi-Missouri (North America)
4) Nile (Africa)
5) Amazon (South America)
6) Yangtze (Chang) (Asia)

Average volume of discharge at mouth, in cubic feet (cubic meters) per second

Rivers with the Largest Drainage Basins

Rank	River	Area of Drainage Basin — Square Miles	Square Kilometers
1	Amazon (Amazonas), South America	2,669,000	6,915,000
2	Congo, Africa	1,474,500	3,820,000
3	Mississippi-Missouri, North America	1,243,000	3,220,000
4	Río de la Plata–Paraná, South America	1,197,000	3,100,000
5	Ob'-Irtysh, Asia	1,154,000	2,990,000
6	Nile, Africa	1,108,000	2,870,000
7	Yenisey-Angara, Asia	1,011,000	2,618,500
8	Lena, Asia	961,000	2,490,000
9	Niger, Africa	807,000	2,090,000
10	Amur-Argun, Asia	792,000	2,051,300
11	Yangtze (Chang), Asia	705,000	1,826,000
12	Volga, Europe	525,000	1,360,000
13	Zambezi, Africa	513,500	1,330,000
14	St. Lawrence, North America	503,000	1,302,800
15	Huang (Yellow), Asia	486,000	1,258,700

Sources for volume and drainage basin information: Atlas of World Water Balance, *USSR National Committee for the International Hydrological Decade and UNESCO, 1977;* Principal Rivers and Lakes of the World, *National Oceanic and Atmospheric Administration, 1982.*

Climate and Weather

Temperature Extremes by Continent

Africa

Highest recorded temperature
Al 'Azīzīyah, Libya, September 13, 1922:
136° F (58° C),
Lowest recorded temperature
Ifrane, Morocco, February 11, 1935:
-11° F (-24° C)

Antarctica

Highest recorded temperature
Vanda Station, January 5, 1974:
59° F (15° C)
Lowest recorded temperature
Vostok, July 21, 1983:
-129° F (-89° C)

Asia

Highest recorded temperature
Tirat Zevi, Israel, June 21, 1942:
129° F (54° C)
Lowest recorded temperature
Oymyakon and Verkhoyansk,
Russia, February 5 and 7, 1892,
and February 6, 1933: -90° F (-68° C)

Australia / Oceania

Highest recorded temperature
Cloncurry, Queensland, January 16, 1889:
128° F (53° C)
Lowest recorded temperature
Charlottes Pass, New South Wales,
June 14, 1945, and July 22, 1947: -8° F (-22° C)

Europe

Highest recorded temperature
Sevilla, Spain, August 4, 1881:
122° F (50° C)
Lowest recorded temperature
Ust' Ščugor, Russia, (date not known):
-67° F (-55° C)

North America

Highest recorded temperature
Death Valley, California, United States,
July 10, 1913: 134° F (57° C)
Lowest recorded temperature
Northice, Greenland, January 9, 1954:
-87° F (-66° C)

South America

Highest recorded temperature
Rivadavia, Argentina, December 11, 1905:
120° F (49° C)
Lowest recorded temperature
Sarmiento, Argentina, June 1, 1907:
-27° F (-33° C)

World

Highest recorded temperature
Al 'Azīzīyah, Libya, September 13, 1922:
136° F (58° C)
Lowest recorded temperature
Vostok, Antarctica, July 21, 1983:
-129° F (-89° C)

World Temperature Extremes

Highest mean annual temperature Dalol, Ethiopia, 94° F (34° C)
Lowest mean annual temperature Plateau Station, Antarctica: -70° F (-57° C)

Greatest difference between highest and lowest recorded temperatures
Verkhoyansk, Russia. The highest temperature ever recorded there is 93.5° F (34.2° C); the lowest is -89.7° F (-67.6° C) — a difference of 183° F (102° C).

Highest temperature ever recorded at the South Pole 7.5° F (-14° C) on December 27, 1978

Most consecutive days with temperatures of 100° F (38° C) or above Marble Bar, Australia, 162 days: October 30, 1923 to April 7, 1924

Greatest rise in temperature within a 12-hour period
Granville, North Dakota, on February 21, 1918. The temperature rose 83° F (46° C), from -33° F (-36° C) in early morning to +50° F (10° C) in late afternoon

Greatest drop in temperature within a 12-hour period
Fairfield, Montana, on December 24, 1924. The temperature dropped 84° F (46° C), from 63° F (17° C) at noon to -21° F (-29° C) by midnight

Temperature Ranges for 14 Major Cities around the World

City	Mean Temperature Coldest Winter Month	Hottest Summer Month	City	Mean Temperature Coldest Winter Month	Hottest Summer Month
Buenos Aires, Argentina	Aug: 51.3° F (10.7° C)	Jan: 75.0° F (23.9° C)	Mumbai (Bombay), India	Jan: 74.3° F (23.5° C)	May: 85.5° F (29.7° C)
Kolkata (Calcutta), India	Jan: 67.5° F (19.7° C)	May: 88.5° F (31.4° C)	New York City, U.S.	Jan: 32.9° F (0.5° C)	Jul: 77.0° F (25.0° C)
London, England	Feb: 39.4° F (4.1° C)	Jul: 63.9° F (17.7° C)	Osaka, Japan	Jan: 40.6° F (4.8° C)	Aug: 82.2° F (27.9° C)
Los Angeles, U.S.	Jan: 56.3° F (13.5° C)	Jul: 74.1° F (23.4° C)	Rio de Janeiro, Brazil	Jul: 70.2° F (21.2° C)	Jan: 79.9° F (26.6° C)
Manila, Philippines	Jan: 77.7° F (25.4° C)	May: 84.9° F (29.4° C)	São Paulo, Brazil	Jul: 58.8° F (14.9° C)	Jan: 71.1° F (21.7° C)
Mexico City, Mexico	Jan: 54.1° F (12.3° C)	May: 64.9° F (18.3° C)	Seoul, South Korea	Jan: 23.2° F (-4.9° C)	Aug: 77.7° F (25.4° C)
Moscow, Russia	Feb: 14.5° F (-9.7° C)	Jul: 65.8° F (18.8° C)	Tokyo, Japan	Jan: 39.6° F (4.2° C)	Aug: 79.3° F (26.3° C)

Precipitation

Greatest local average annual rainfall
Mt. Waialeale, Kaua'i, Hawaii,
460 inches (1,168 cm)

Lowest local average annual rainfall
Arica, Chile, .03 inches (.08 cm)

Greatest rainfall in 12 months
Cherrapunji, India, August 1860 to August 1861:
1,042 inches (2,647 cm)

Greatest rainfall in one month
Cherrapunji, India, July 1861: 366 inches (930 cm)

Greatest rainfall in 24 hours
Cilaos, Reunion, March 15 and 16, 1952:
74 inches (188 cm)

Greatest rainfall in 12 hours
Belouve, Reunion, February 28 and 29, 1964:
53 inches (135 cm)

Most thunderstorms annually
Kampala, Uganda, averages 242 days per
year with thunderstorms

Between 1916 and 1920, Bogor, Indonesia,
averaged 322 days per year with thunderstorms

Longest dry period
Arica, Chile, October, 1903
to January, 1918 — over 14 years

Largest hailstone ever recorded
Coffeyville, Kansas, U.S., September 3, 1970:
circumference 17.5 inches (44.5 cm)
diameter 5.6 inches (14 cm),
weight 1.67 pounds (758 grams)

Heaviest hailstone ever recorded
Kazakhstan, 1959: 4.18 pounds (1.9 kilograms)

North America's greatest snowfall in one season
Rainier Paradise Ranger Station, Washington,
U.S., 1971–1972: 1,122 inches (2,850 cm)

North America's greatest snowfall in one storm
Mt. Shasta Ski Bowl, California, U.S.,
February 13 to 19, 1959: 189 inches (480 cm)

North America's greatest snowfall in 24 hours
Silver Lake, Colorado, U.S., April 14 and 15, 1921:
76 inches (1 92.5 cm)

N. America's greatest depth of snowfall on the ground
Tamarack, California, U.S., March 11, 1911:
451 inches (1,145.5 cm)

Foggiest place on the U.S. West Coast
Cape Disappointment, Washington,
averages 2,552 hours of fog per year

Foggiest place on the U.S. East Coast
Mistake Island, Maine, averages
1,580 hours of fog per year

Wind

Highest 24-hour mean surface wind speed
Mt. Washington, New Hampshire, U.S.,
April 11 and 12, 1934: 128 mph (206 kph)

Highest 5-minute mean surface wind speed
Mt. Washington, New Hampshire, U.S.,
April 12, 1934: 188 mph (303 kph)

Highest surface wind peak gust:
Mt. Washington, New Hampshire, U.S.,
April 12, 1934: 231 mph (372 kph)

Windiest U.S. Cities

Chicago is sometimes called The Windy City.
It earned this nickname because of long-winded politicians,
not because it has the strongest gales.

The windiest cities in the U.S. are as follows:

Cities	Average wind speed	
	mph	kph
Great Falls, Montana	13.1	21.0
Oklahoma City, Oklahoma	13.0	20.9
Boston, Massachusetts	12.9	20.7
Cheyenne, Wyoming	12.8	20.6
Wichita, Kansas	12.7	20.4

Chicago ranks 16th, with a 10.4 mph (16.7 kph) average.

Deadliest Hurricanes in the U.S. since 1890

Rank	Place	Year	Number of Deaths
1	Texas (Galveston)	1900	8,000
2	Louisiana	1893	2,000
3	Florida (Lake Okeechobee)	1928	1,836
4	South Carolina, Georgia	1893	>1,000
5	Florida (Keys)	1919	>600
6	New England	1938	600
7	Florida (Keys)	1935	408
8	Southwest Louisiana, north Texas—"Hurricane Audrey"	1957	390
	Northeast U.S.	1944	390
9	Louisiana (Grand Isle)	1909	350
10	Louisana (New Orleans)	1915	275

Tornadoes in the U.S., 1950—1993

Rank	State	Total Number of Tornadoes	Yearly Average	Total Number of Deaths
1	Texas	5,303	120	471
2	Oklahoma	2,259	51	217
3	Kansas	2,068	47	199
4	Florida	1,932	44	81
5	Nebraska	1,618	37	51
	U.S. Total	33,120	753	4,045

Deadliest Floods in the U.S. since 1900

Rank	Place	Year	Number of Deaths
1	Ohio River and tributaries	1913	467
2	Mississippi Valley	1927	313
3	Black Hills, South Dakota	1972	237
4	Willow Creek, Oregon	1903	225
5	Texas rivers	1921	215
6	Northeastern U.S., following Hurricane Dianne	1955	187
7	Texas rivers	1913	177
8	James River basin, Virginia	1969	153
9	New England	1936	150+
10	Big Thompson Canyon, Colorado	1976	139
11	Ohio and Lower Mississippi river basins	1937	137
12	Buffalo Creek, West Virginia	1972	125

Population

During the first two million years of our species' existence, human population grew at a very slow rate, and probably never exceeded 10 million. With the development of agriculture circa 8000 B.C., the growth rate began to rise sharply: by the year A.D. 1, the world population stood at approximately 250 million.

By 1650 the population had doubled to 550 million, and within only 200 years it doubled again, reaching almost 1.2 billion by 1850. Each subsequent doubling has taken only about half as long as the previous one: 100 years to reach 2.5 billion, and 40 years to reach 5.2 billion.

World Population

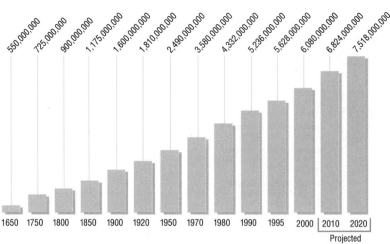

| 550,000,000 | 725,000,000 | 900,000,000 | 1,175,000,000 | 1,600,000,000 | 1,810,000,000 | 2,490,000,000 | 3,580,000,000 | 4,332,000,000 | 5,236,000,000 | 5,628,000,000 | 6,080,000,000 | 6,824,000,000 | 7,518,000,000 |

| 1650 | 1750 | 1800 | 1850 | 1900 | 1920 | 1950 | 1970 | 1980 | 1990 | 1995 | 2000 | 2010 | 2020 |

Projected Population

Historical Populations of the Continents and the World

Year	Africa	Asia	Australia	Europe	North America	Oceania, incl. Australia	South America	World
1650	*100,000,000*	335,000,000	*<1,000,000*	*100,000,000*	*5,000,000*	*2,000,000*	*8,000,000*	*550,000,000*
1750	*95,000,000*	476,000,000	*<1,000,000*	*140,000,000*	*5,000,000*	*2,000,000*	*7,000,000*	*725,000,000*
1800	*90,000,000*	593,000,000	*<1,000,000*	*190,000,000*	*13,000,000*	*2,000,000*	*12,000,000*	*900,000,000*
1850	*95,000,000*	754,000,000	*<1,000,000*	265,000,000	*39,000,000*	*2,000,000*	*20,000,000*	*1,175,000,000*
1900	*118,000,000*	932,000,000	4,000,000	400,000,000	106,000,000	6,000,000	38,000,000	*1,600,000,000*
1920	*140,000,000*	1,000,000,000	6,000,000	453,000,000	147,000,000	9,000,000	61,000,000	*1,810,000,000*
1950	199,000,000	1,418,000,000	8,000,000	530,000,000	219,000,000	13,000,000	111,000,000	*2,490,000,000*
1970	346,900,000	2,086,200,000	12,460,000	623,700,000	316,600,000	19,200,000	187,400,000	3,580,000,000
1980	463,800,000	2,581,000,000	14,510,000	660,000,000	365,000,000	22,700,000	239,000,000	4,332,000,000
1990	648,300,000	3,156,100,000	16,950,000	688,000,000	423,600,000	26,300,000	293,700,000	5,236,000,000

Figures for years prior to 1970 are rounded to the nearest million. Figures in italics represent rough estimates.

The Most Populous City in the World, through History

With more than 30 million people, Japan's Tokyo-Yokohama agglomeration ranks as the most populous metropolitan area in the world today. New York City held this title from the mid-1920's through the mid-1960's. But what city was the most populous in the world five hundred years ago? Five *thousand* years ago?

The following time line represents one expert's attempt to name the cities that have reigned as the most populous in the world since 3200 B.C. The time line begins with Memphis, the capital of ancient Egypt, which was possibly the first city in the world to attain a population of 20,000.

Listed after each city name is the name of the political entity to which the city belonged during the time that it was the most populous city in the world. The name of the modern political entity in which the city, its ruins, or its site is located, where this entity differs from the historic political entity, is listed in parentheses.

For the purpose of this time line, the word "city" is used in the general sense to denote a city, metropolitan area, or urban agglomeration.

It is important to note that reliable census figures are not available for most of the 5,200 years covered by this time line. Therefore the time line is somewhat subjective and conjectural.

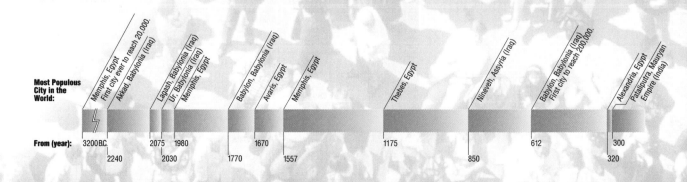

Most Populous City in the World:

Memphis, Egypt First city ever to reach 20,000.
Akkad, Babylonia (Iraq)
Lagash, Babylonia (Iraq)
Ur, Babylonia (Iraq)
Memphis, Egypt
Babylon, Babylonia (Iraq)
Avaris, Egypt
Memphis, Egypt
Thebes, Egypt
Nineveh, Assyria (Iraq)
Babylon, Babylonia (Iraq) First city to reach 200,000.
Alexandria, Egypt
Pataliputra, Mauryan Empire (India)

From (year): 3200 B.C.
2240
2075
1980
2030
1670
1770
1557
1175
850
612
300
320

Most Densely Populated Countries

Rank	Country (Population)	Population per Square Mile	Population per Square Kilometer
1	Monaco (32,000)	45,489	16,759
2	Singapore (4,375,000)	17,481	6,762
3	Vatican City (1,000)	5,000	2,500
4	Malta (395,000)	3,234	1,249
5	Maldives (315,000)	2,702	1,043
6	Bahrain (650,000)	2,417	934
7	Bangladesh (132,315,000)	2,361	912
8	Guernsey (64,000)	2,144	825
9	Jersey (89,000)	1,986	770
10	Barbados (275,000)	1,659	640
11	Taiwan (22,460,000)	1,609	621
12	Mauritius (1,195,000)	1,510	583
13	Nauru (12,000)	1,492	576
14	Korea, South (48,120,000)	1,253	484
15	San Marino (28,000)	1,136	448

Least Densely Populated Countries

Rank	Country (Population)	Population per Square Mile	Population per Square Kilometer
1	Greenland (56,000)	0.07	0.03
2	Mongolia (2,675,000)	4.4	1.7
3	Namibia (1,810,000)	5.6	2.2
4	Australia (19,455,000)	6.5	2.5
5	Mauritania (2,790,000)	6.9	2.6
6	Suriname (435,000)	6.9	2.6
7	Iceland (280,000)	7.0	2.7
8	Botswana (1,590,000)	7.1	2.7
9	Libya (5,305,000)	7.7	3.0
10	Canada (31,750,000)	8.2	3.2
11	Guyana (695,000)	8.4	3.2
12	Gabon (1,225,000)	11.8	4.6
13	Central African Republic (3,610,000)	14.9	5.7
14	Kazakhstan (16,735,000)	15.9	6.2
15	Chad (8,850,000)	17.6	6.8

Most Highly Urbanized Countries

Country	Urban pop. as a % of total pop.
Vatican City	100%
Singapore	100%
Monaco	100%
Belgium	96%
Kuwait	96%
Iceland	92%
Uruguay	91%
Israel (excl. Occupied Areas)	90%
Qatar	90%
Andorra	89%
Argentina	89%
San Marino	89%
United Kingdom	89%
Bahrain	88%
Malta	87%

Least Urbanized Countries

Country	Urban pop. as a % of total pop.
Bhutan	5%
Burundi	5%
Rwanda	5%
Nepal	9%
Oman	11%
Cambodia (Kampuchea)	14%
Bangladesh	14%
Uganda	15%
Burkina Faso	15%
Eritrea	15%
Grenada	15%
Solomon Islands	15%
Niger	15%
Ethiopia	16%
Nigeria	16%

World's Largest Metropolitan Areas

Rank	Name	Population
1	Tōkyō-Yokohama, Japan	31,915,000
2	Seoul, South Korea	21,450,000
3	Mexico City, Mexico	20,150,000
4	New York City, U.S.	19,500,000
5	Jakarta, Indonesia	17,600,000
6	São Paulo, Brazil	17,480,000
7	Ōsaka-Kōbe-Kyōto, Japan	17,350,000
8	Mumbai (Bombay), India	16,600,000
9	Delhi-New Delhi, India	16,000,000
10	Los Angeles, U.S.	15,200,000
11	Kolkata (Calcutta), India	14,000,000
12	Buenos Aires, Argentina	13,900,000
13	Cairo, Egypt	13,300,000
14	Moscow, Russia	12,800,000
15	London, England	12,700,000

The 50 Most Populous Countries

Rank	Country	Population	Rank	Country	Population	Rank	Country	Population
1	China	1,278,720,000	18	Iran	66,365,000	35	Canada	31,750,000
2	India	1,037,955,000	19	Thailand	62,080,000	36	Kenya	30,960,000
3	United States	279,310,000	20	United Kingdom	59,715,000	37	Morocco	30,905,000
4	Indonesia	230,260,000	21	France	59,660,000	38	Peru	27,720,000
5	Brazil	175,260,000	22	Italy	57,700,000	39	Afghanistan	27,280,000
6	Pakistan	146,145,000	23	Dem. Rep. of the Congo	54,455,000	40	Nepal	25,580,000
7	Russia	145,215,000	24	Ukraine	48,570,000	41	Uzbekistan	25,355,000
8	Bangladesh	132,315,000	25	Korea, South	48,120,000	42	Uganda	24,335,000
9	Nigeria	128,285,000	26	South Africa	43,645,000	43	Venezuela	24,105,000
10	Japan	126,880,000	27	Myanmar	42,120,000	44	Iraq	23,665,000
11	Mexico	102,640,000	28	Colombia	40,680,000	45	Saudi Arabia	23,130,000
12	Philippines	83,685,000	29	Spain	40,060,000	46	Taiwan	22,460,000
13	Germany	83,145,000	30	Poland	38,630,000	47	Malaysia	22,445,000
14	Vietnam	80,520,000	31	Argentina	37,600,000	48	Romania	22,340,000
15	Egypt	70,125,000	32	Tanzania	36,705,000	49	Korea, North	22,100,000
16	Turkey	66,905,000	33	Sudan	36,585,000	50	Ghana	20,070,000
17	Ethiopia	66,780,000	34	Algeria	32,005,000			

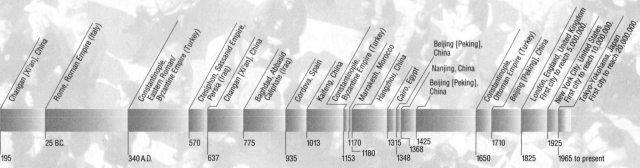

Changan [Xi'an], China
Rome, Roman Empire (Italy)
Constantinople, Eastern Roman/ Byzantine Empire (Turkey)
Ctesiphon, Sassanid Empire, Persia (Iraq)
Changan [Xi'an], China
Baghdad, Abbasid Caliphate (Iraq)
Cordova, Spain
Kaifeng, China
Constantinople, Byzantine Empire (Turkey)
Marrakesh, Morocco
Hangzhou, China
Cairo, Egypt
Beijing [Peking], China
Nanjing, China
Beijing [Peking], China
Constantinople, Ottoman Empire (Turkey)
Beijing [Peking], China
London, England, United Kingdom. First city to reach 5,000,000.
New York City, United States. First city to reach 10,000,000.
Tokyo-Yokohama, Japan. First city to reach 20,000,000.

195
25 B.C.
340 A.D.
570
637
775
935
1013
1153
1170
1180
1315
1348
1368
1425
1650
1710
1825
1925
1965 to present

Source: Four Thousand Years of Urban Growth by Tertius Chandler, Edwin Mellen Press, 1987.

Economics and Energy

Cattle

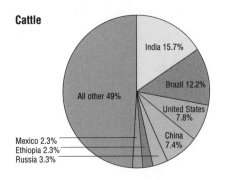

Total annual world production: 1,318,408,000 head

Hogs

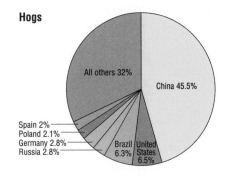

Total annual world production: 902,866,000 head

Corn

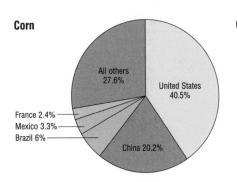

Total annual world production: 559,077,000 metric tons

Crude Steel

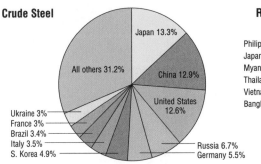

Total annual world production: 747,333,000 metric tons

Rice

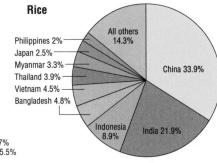

Total annual world production: 553,032,000 metric tons

Gross Domestic Product

Annual gross domestic product is the total market value of all the goods and services produced by a nation in a year. Annual per capita gross domestic product is the total value of the goods and services divided by the nation's population.

The United States, Canada, Chile, the countries of western Europe, Australia, Japan, South Korea, Malaysia, and Taiwan have annual per capita GDP's of more than $10,000 U.S. dollars. Most of the countries of South America have annual per capita GDP's between $5,000 and $10,000. Most of the countries of Asia and eastern Europe have annual per capita GDP's between $1,000 and $5,000. Most African countries have annual per capita GDP's under $2,500.

Percentage of World Population in each Per Capita GDP Category

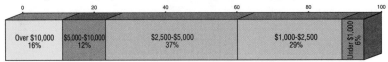

Electricity

Hydroelectricity

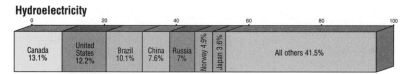

Total annual world production: 2,533,000 gigawatt hours

Nuclear Energy

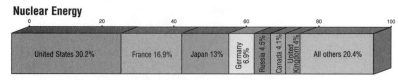

Total annual world production: 2,268,000 gigawatt hours

World Electricity Production

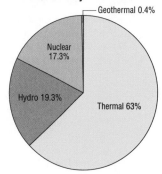

Total annual world electricity production:
13,098,000 gigawatt hours

Petroleum

Petroleum Production

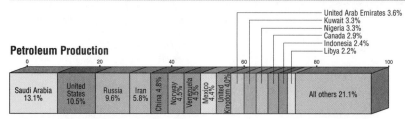

Total annual world production: 22,797,544,000 barrels

Petroleum Reserves

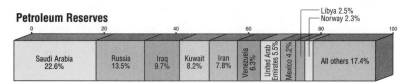

Total world reserves: 1,160,069,500,000 barrels

Commercial Energy

Commercial Energy Production

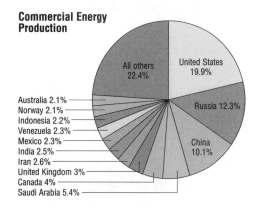

Total annual world production: equivalent of 12,321,830,000
metric tons of coal

Commercial Energy Consumption

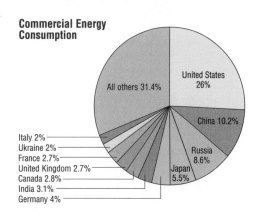

Total annual world consumption: equivalent of 11,720,193,000
metric tons of coal

Questions and Answers

North America

Q What is the difference between "Central America" and "Middle America"?

A The term "Central America" refers to the North American countries which lie south of Mexico and north of Colombia: Belize, Guatemala, Honduras, El Salvador, Nicaragua, Costa Rica, and Panama. The Caribbean islands are not considered part of Central America. "Middle America" is comprised of Central America as well as Mexico and all of the Caribbean islands.

Q What is the largest U.S. city east of Reno, Nevada and west of Chicago?

Los Angeles, California

A Surprisingly, the answer is Los Angeles. Although Los Angeles is located on the Pacific Coast, it actually lies slightly farther east than Reno. The coast, which forms the western edge of the U.S., curves dramatically eastward below Cape Mendocino in northern California. San Diego, at the southern end of California's coast, is approximately as far east as the eastern borders of Washington and Oregon. (*Refer to map on pages 74-75.*)

Q Which is the southernmost U.S. state? The northernmost? The westernmost? The easternmost?

A Hawaii is the southernmost state; Alaska is both the westernmost and the northernmost. The question of which state is the easternmost is a bit problematic. Generally, Maine is considered to be the easternmost, since it extends farther east than any other state along the Atlantic seaboard. However, Alaska is technically the easternmost state, since the Aleutian Islands cross the 180° longitude line which divides the globe into eastern and western hemispheres. These islands sit at the eastern edge of the eastern hemisphere.

Q How many national flags have flown over Texas?

A Six. Spain (1682-1821), France (1685-1686), Mexico (1821-1836), the Republic of Texas (1836-1845), the United States (1845-1861), the Confederate States of America (from 1861 until the state was re-admitted to the Union in 1870), and the United States again (1870 through the present). Texas is the only U.S. state to have existed as an independent country.

Q What is the oldest city in the United States to be founded by Europeans?

A St. Augustine, Florida. Spanish explorer Juan Ponce de León, searching for the Fountain of Youth, landed nearby and claimed the area for Spain in 1513. The French established a colony on the site in 1564, but it was destroyed in 1565 by the Spanish, who then founded the present city. The oldest U.S. state capital is Santa Fe, New Mexico, which was founded in 1609, also by the Spanish.

Q What is Papiamento?

A Papiamento is a language which blends Dutch, Spanish, Portuguese, English, and Indian words. It is the principal language of Aruba and other islands in the Dutch Caribbean, and is spoken by an estimated 200,000 people.

Q If you were on a ship sailing from the Atlantic Ocean to the Pacific Ocean, in which direction would you be traveling as you passed through the Panama Canal?

A Southeast. The Pacific lies west of the Atlantic, and it would seem that a ship passing through the canal from the Atlantic would be sailing west. However, because of the twisting shape of the Isthmus of Panama, the canal's Pacific end lies south and east of its Atlantic end. (*Refer to inset map on page 96.*)

Q Into what body of water does the Colorado River empty?

A Currently, it doesn't empty into any body of water. Until recently, the river flowed into the Gulf of California. As the populations of water-poor Arizona and California have soared, more and more water has been drawn from the river for farms, industry, and homes. Today the Colorado is barely a trickle when it crosses the border into Mexico, and it disappears in the desert before it reaches the Gulf of California.

Q What is Canada's smallest province? Its largest?

A Prince Edward Island is Canada's smallest province, at 2,185 square miles (5,660 sq km). Canada's largest province is Quebec, which covers 594,860 square miles (1,540,680 sq km). The newly created territory of Nunavut represents the country's largest administrative division: It spreads over an area of 733,594 square miles (1,900,000 sq km), much of which lies within the Arctic Circle. If Nuvavut were an independent country, it would be larger than all but 15 of the world's countries.

Q What is the Continental Divide?

A An imaginary line running down the backbone of North America. Except for those which empty into the Great Basin and other basins, rivers to the west of this line flow into the Pacific Ocean, including its bays and gulfs; rivers to the east flow into the Atlantic or Arctic oceans, including their bays and gulfs. From northwest Canada south to New Mexico, the Divide runs along the crest of the Rocky Mountains, and in northern Mexico it follows the ridge of the Sierra Madre Occidental. All continents except frozen Antarctica have "divides."

Q Is Niagara Falls the highest waterfall in the world?

A Not even close. Niagara Falls' maximum drop of 167 feet (51 m) is surpassed by at least 22 waterfalls in North America alone. The highest waterfall in the world is Angel Falls in Venezuela, which spills 3,212 feet (979 m) from a flat mountain plateau. However, Niagara Falls ranks first in a different category: more than 210,000 cubic feet of water cascade over its edge each second, an amount nearly double that of any other waterfall.

Niagara Falls actually consists of two separate waterfalls: American Falls, at left, and Canadian (Horseshoe) Falls.

Q What did ships do before the Panama Canal was built?

A Before the canal opened, ships traveling between New York and San Francisco would sail 13,000 miles (21,000 km) around the entire continent of South America. When the canal opened in 1914, this journey was shortened to 5,200 miles (8,400 km). However, the canal is now too narrow for many of today's largest ocean-going vessels, so they must once again sail around South America to reach their destination. A project to widen the canal began in 1992.

Gatun Lake, Panama Canal

Q What is the largest inland body of water in Central America?

A Lake Nicaragua, which has a surface area of 3,150 square miles (8,158 sq km). Its only outlet is the San Juan River, which flows into the Caribbean Sea. Lying in a lowland region called the Nicaragua Depression, the lake was once part of the sea but became separated when the land began to rise. The freshwater lake is home to many species of fish usually found only in salt water, including sharks, tuna, and swordfish.

Q Why does Minnesota have so many lakes?

A During the height of the most recent Ice Age, glaciers moved southward from the Arctic regions to cover Canada and much of the northern U.S., including Minnesota. As they advanced, the glaciers scoured the landscape, gouging out countless depressions. When the Earth's climate grew warmer, the glaciers began to melt and retreat, and the depressions filled with meltwater to become lakes. Although Minnesota bills itself as the "Land of 10,000 Lakes," it actually has more than 15,000 lakes. Other areas of the world which experienced extreme glaciation, including parts of Europe and Siberia, also contain many lakes.

Q What is the most popular U.S. national park?

A Great Smoky Mountains National Park. Covering over 521,500 acres (211,200 hectares) in eastern Tennessee, the park receives approximately 9.1 million visitors each year. Arizona's Grand Canyon National Park has the second-highest visitor count: it receives around 4.6 million people annually.

Q What two U.S. states share borders with the most other states?

A Missouri and Tennessee, which each border eight other states. Missouri borders Iowa, Illinois, Kentucky, Tennessee, Arkansas, Oklahoma, Kansas, and Nebraska. Tennessee borders Kentucky, Virginia, North Carolina, Georgia, Alabama, Mississippi, Arkansas, and Missouri.

Q How many U.S. states border only one other state?

A One: Maine, which borders only New Hampshire. Two states, Alaska and Hawaii, border no others.

Q What is the one place in North America where you can see both the Atlantic Ocean and the Pacific Ocean?

A Irazú, a volcano in central Costa Rica. From its summit, both the Atlantic and Pacific oceans can be seen on a clear day.

Q Where is the Yucatan Peninsula?

A This thumb of land juts off of southeastern Mexico, separating the Gulf of Mexico from the Caribbean Sea. Yucatan was the center of the Maya civilization from about the first century B.C. through the tenth century A.D. Extensive Mayan ruins can be found at Chichén Itzá, Kabah, Mayapán, Tulum, and Uxmal. Near the town of Chicxulub at the northern tip of the peninsula, there is evidence of an enormous crater which is thought to be the point of impact of a meteorite 65 million years ago. The impact would have sent so much dust into the atmosphere that the sun's rays would have been blocked for months or perhaps years, lowering temperatures globally and possibly causing the extinction of the dinosaurs.

Q What is the oldest capital city in the Americas?

A Mexico City. The city originated as Tenochtitlán, the capital of the Aztecs, founded in the mid-1300s. By the early 1500s, the city had a population of perhaps 150,000, which was not only greater than any other city in the Americas but also greater than any European city at the time. In 1521, after a three-month siege, Spanish invaders under Hernán Cortés captured Tenochtitlán, razed the entire city, and founded Mexico City upon its ruins.

Q What is the most densely populated country in North America?

A El Salvador, which has a density of 775 people per square mile (299 per sq km). The U.S. ranks eighth, with 75 people per square mile (29 per sq km). Canada, the continent's largest country, is by far the least densely populated: it averages only 8.2 people per square mile (3.2 per sq km).

Q What is the largest island in Caribbean Sea?

A Cuba, the world's 15th-largest island. It has a land area of 42,800 square miles (110,800 sq km). The second-largest Caribbean island is Hispaniola, which covers 29,400 square miles (76,200 sq km) and contains the countries of Haiti and the Dominican Republic. Jamaica, measuring 4,200 square miles (11,000 sq km), is the third-largest Caribbean island.

Q How many U.S. states have volcanoes that were active in the 20th century?

A Four. Alaska has the most: 34 of its volcanoes, most of which are located on the Alaska Peninsula and the Aleutian Islands, have erupted since 1900. The other states with documented eruptions in the 20th century are: Hawaii (3), Washington (1), and California (1).

Caldera and lava lake of Kīlauea, on the island of Hawaiʻi

Q What is the only Caribbean island with large oil reserves?

A Trinidad. The island's economy is based on oil, which accounts for about 80% of its exports. As a result of oil wealth, Trinidadians enjoy a higher standard of living than the people of most other Caribbean countries.

Q Which U.S. state has the highest average elevation?

A Colorado, with an average elevation of 6,800 feet (2,074 m) above sea level. However, the four highest peaks in the U.S. are found not in Colorado but in Alaska.

Questions and Answers

South America

Q What is Latin America?

A The term "Latin America" designates the parts of North and South America which were settled by Spanish and Portuguese colonists and still retain a Hispanic character. These include Mexico, Cuba, Puerto Rico, the Dominican Republic, some of the smaller islands in the West Indies, all of Central America except for Belize, and all of South America except for Guyana, Suriname, and French Guiana.

Q If you flew due south from Chicago, which South American country would you fly over first?

A You wouldn't fly over any South American countries. A straight line drawn south from Chicago passes through the Gulf of Mexico, Central America, and the Pacific Ocean. Point Parinas, Peru, the westernmost point of mainland South America, is about 500 miles (800 km) east of this line. The Galapagos Islands, which belong to Ecuador, lie about 75 miles (120 km) west of the line.

Q What percentage of the world's coffee beans come from South America?

A South America currently produces approximately 45% of the world's coffee beans. Brazil leads the continent, producing just under one-quarter of the world total. Another 16% are grown in Colombia. Coffee plants require hot, moist climates, and they yield the most flavorful beans when cultivated at elevations between 3,000 and 6,000 feet (900 and 1800 m). South America's principal coffee-growing regions are found in the Brazilian and Guiana Highlands, and in the valleys and foothills of the Andes.

Giant Galapagos turtle

Q What scientist made the Galapagos Islands famous?

A Charles Darwin, who visited the islands during his 1831-1836 expedition on the H.M.S. *Beagle.* Darwin's observations of how various animal species had adapted to life on the islands contributed to his ground-breaking theory of evolution, which he presented in the 1859 book *On the Origin of Species.*

Q Where are the Falkland Islands?

A In the Atlantic Ocean, about 275 miles (440 km) off the east coast of Argentina. The Falklands are a dependency of the United Kingdom, and nearly all of the residents are English-speakers of British descent. Argentina, which has asserted claims to the Falklands since 1816, invaded and occupied the islands in 1982. The U.K. sent a large task force that defeated the Argentineans in a war lasting less than a month.

Q What South American country is the longest when measured from north to south?

A Brazil. The country measures 2,725 miles (4,395 km) north to south, which is the approximate distance from New York City to Reno, Nevada. The second-longest country is Chile, with a length of 2,647 miles (4,270 km). In contrast to its great length, Chile measures only 235 miles (380 km) east to west at its widest point.

Q What was discovered in Venezuela's Lake Maracaibo in 1914?

A Oil. Maracaibo sits above one of the world's largest oil fields, and today the lake's surface is a thicket of oil derricks. Wealth from oil exportation has helped to make Venezuela one of the richest countries in Latin America.

Q What is Patagonia?

A Patagonia is a wind-swept plateau region occupying the southern third of South America, east of the Andes. The plateau receives little precipitation, and its only vegetation is scrubby grasses and thorny desert shrubs. The name "Patagonia" probably comes from the Grand Patagon, a dog-headed monster in a European romance called *Primaleon of Greece.* In 1592, the crew members of the ship *Desire* were attacked by a war-party of Tehuelche Indians wearing dog masks.

Q What South American city is the world's highest national capital?

A La Paz, Bolivia. The city sprawls across the floor of a deep canyon high in the Andes, at an elevation of 12,000 feet (19,350 km)—approximately the same height as the summit of Japan's Mt. Fuji. The canyon walls protect the city from the bitterly cold winds that whip across the surrounding plateau. Visitors from lower elevations often suffer from altitude sickness for several days until their bodies adjust to the thin, oxygen-poor air.

Q What is the southernmost city in the world?

A Ushuaia, Argentina, on the island of Tierra del Fuego. This city of 29,452 people lies at 54°48' south latitude, less than 700 miles (1,100 km) from Antarctica. The world's southernmost city with a population greater than 100,000 is Punta Arenas, Chile, located 150 miles (240 km) northwest of Ushuaia.

Q How many South American countries are named for famous people or cities?

A Three. Bolivia was named for Simón Bolívar, the revolutionary who helped to liberate much of northern South America from Spanish rule. Colombia takes its name from the explorer Christopher Columbus, who "discovered" South America in 1498 and sailed along the coast of present-day Colombia in 1502. The name "Venezuela" is Spanish for "Little Venice." When European explorers reached Lake Maracaibo in 1499, they found villages built on pilings over the shallow waters, which reminded them of Venice, Italy.

Q Why is South America sometimes referred to as "the hollow continent"?

A South America earned this nickname because most of its people live on or near the coasts; the interior is sparsely populated. Sixteen of the 20 largest metropolitan areas lie within 200 miles (320 km) of the coast. The continent's uneven population distribution has both a historical and a geographical explanation. Beginning in the 16th century, European colonists settled in the coastal regions from which raw materials were shipped back to Europe. The Amazon rain forest—hot, humid, and nearly impenetrable in places—has discouraged settlement of the northern half of the continent, and the Andes present a formidable barrier to eastward expansion from the Pacific coast. *(Refer to population map on page 99.)*

How many of South America's 20 longest rivers empty into the Pacific Ocean?

None. South America's continental divide runs along the crest of the Andes at the continent's western edge. In most places the divide lies within 200 miles (320 km) of the Pacific coast. Rivers originating west of the divide have only a short distance to travel before reaching the Pacific. Rivers flowing east from the divide have nearly the whole expanse of the continent to cross before emptying into the Atlantic.

What South American city was the capital of the Inca empire?

Cathedral and rooftops in Cusco

Cusco, Peru. The city was built by the Incas in the fourteenth century and served as their capital for 200 years until it was destroyed by Francisco Pizarro in 1533. Today, Cusco thrives as a major tourist attraction. Many of its houses and buildings are constructed on foundations of stone first cut by the Incas. Lying about 50 miles (80 km) northwest of Cusco is Machu Picchu, a well-preserved mountaintop Inca city, which was rediscovered in 1911.

What stretch of ocean off the South American coast is considered one of the most treacherous to ships?

The Drake Passage. Approximately 500 miles (1,800 km) wide, this strait separates Cape Horn—at the southern tip of South America—from the South Shetland Islands which lie just north of Antarctica. First traversed in 1615, the passage was part of a major trade route between the Atlantic and Pacific oceans until 1914, when the Panama Canal opened. Frigid temperatures, rough waters, and high winds make the passage treacherous for all vessels but especially for the sailing ships of centuries past. Because the passage is so perilous, many ships avoid it by cutting through the Strait of Magellan to the north. However, this strait has its own dangers: it is narrow and twisting, and has been the site of numerous major shipwrecks.

Where is the Land of Fire?

At the southern tip of South America. Tierra del Fuego, Spanish for "Land of Fire," is a group of islands lying south of the Strait of Magellan. When Portuguese explorer Ferdinand Magellan arrived in 1520, he named the islands after observing the native inhabitants carrying torches. The largest island, also called Tierra del Fuego, accounts for two-thirds of the land area of the island group. The eastern third of the islands belong to Argentina, and the rest belong to Chile.

What are the Nazca Lines?

The Nazca Lines are gigantic drawings that were etched into Peru's desert floor by the Nazca people between 500 B.C. and A.D. 500. Scattered across 200 square miles (520 sq km), they form the world's largest display of art. The drawings fall into two categories: animal motifs and geometric patterns of crisscrossing straight lines. They were made by scraping away the top layer of red gravel to reveal the yellow sand below it. Archaeologists still disagree about the purpose of the Nazca Lines, but some theories hold that the lines formed ancient highways or were used as a giant calendar.

What is "El Niño"?

El Niño is a seasonal ocean current that flows south along the Pacific coast of South America. The current takes its name—which is Spanish for "the Child"—from its usual arrival during the Christmas season. In normal years, when El Niño reaches northern Peru, southeasterly trade winds push its warm surface waters westward across the Pacific, away from the coast. Every four or five years, these trade winds weaken, allowing El Niño to travel farther south along the coast, raising local water temperatures by several degrees. This warmer water kills plankton and fish, crippling the fishing industry. Increased evaporation leads to excessive rainfall over parts of South America. The change in El Niño's normal flow pattern affects other ocean currents which often leads to dramatic climatic changes around the world.

How many of the Earth's species are found in the Amazon rain forest?

Most of the Amazon basin has not been fully explored, and therefore most of its plant and animal species have not yet been catalogued. However, scientists estimate that the Amazon rain forest, which covers less than 5% of the Earth's total land area, contains almost one-half of the planet's animal and plant species. One in ten of the most common medicines we use today comes from rain forest plants, and scientists believe that cures for many diseases, such as cancer, might be derived from plant species not yet discovered.

Where in South America could you find places in which no rainfall has ever been recorded?

The Atacama Desert in northern Chile. This barren land of sand, rocks, borax lakes and saline deposits is one of the driest regions in the world. In parts of the desert, no rainfall has ever been recorded, and the city of Arica, at the northern edge, endured more than 14 consecutive years of drought from October 1903 through January 1918.

What South American possession lies farthest from the mainland?

Ancient statues, Easter Island

Easter Island, situated 2,300 miles (3,700 km) west of Chile in the Pacific Ocean. This small volcanic island was discovered on Easter Sunday in 1722, and was annexed by Chile in 1888. Although it belongs to Chile, geographically Easter Island is considered to be part of Oceania, not South America. It is best known for its strange monuments: scattered over the island are more than 600 huge stone faces, the earliest dating back more than 1,500 years.

Where are the pampas?

These flat, grassy plains—which are much like the prairies of North America—are found in the temperate regions of southern South America, east of the Andes. The largest such plain, known simply as the Pampa, covers much of central and northern Argentina, and extends into Uruguay. Since the 1550s, when European colonists introduced cattle to the Pampa, livestock raising has been a thriving industry. For many people, gauchos, or Argentinean cowboys, are the enduring symbol of the Pampa, although in the last century farming has superseded cattle ranching in economic importance.

Questions and Answers

Europe

Where is the Black Forest?

In southwestern Germany, between the Rhine and Neckar rivers. The Black Forest, or *Schwarzwald* in German, is a mountainous region that takes its name from the dark coniferous trees that cover its slopes. Its fertile valleys provide good pastureland and produce grapes for wine, and its trees supply the lumber and woodworking industries, as well as toy and cuckoo clock manufacturers. The region's scenic beauty, winter sports facilities, and mineral springs attract many tourists each year.

How many national capitals are located on the Danube River?

Old Town of Zagreb, Croatia

Four. The capital cities of Bratislava (Slovakia), Budapest (Hungary), Belgrade (Serbia and Montenegro), and Vienna (Austria) are all found along the banks of the Danube. Five other capitals are located on tributaries of the Danube: Bucharest (Romania), Sofia (Bulgaria), Ljubljana (Slovenia), Zagreb (Croatia), and Sarajevo (Bosnia and Herzegovina).

What is killing the forests of Northern Europe?

Acid rain. In the atmosphere, airborne pollutants—especially sulfur and nitrogen dioxides from automobile and industrial emissions—adhere to water droplets, and then fall back to Earth as acidified rain, snow, or hail. This precipitation poisons plant and animal life, erodes buildings, and contaminates soil and drinking water. As a result of acid rain, as many as one-half of the trees in Germany's Black Forest and Switzerland's central alpine region are dead or dying. At least 4,000 lakes in Sweden are so acidic that no fish survive in them. To combat acid rain, the countries of the European Union recently agreed to significantly reduce nitrogen oxide and sulfur dioxide emissions.

What independent countries were once part of the U.S.S.R.?

When the Union of Soviet Socialist Republics (U.S.S.R.) broke up in 1991, its 15 republics all became independent countries: Armenia, Azerbaijan, Belarus, Estonia, Georgia, Kazakhstan, Krygyzstan, Latvia, Lithuania, Moldova, Russia, Tajikistan, Turkmenistan, Ukraine, and Uzbekistan.

How many times did the name of St. Petersburg, Russia, change in the 20th century?

Three times. St. Petersburg was founded in 1703 by Peter the Great. In 1914, its name was changed to Petrograd, Russian for "Peter's City," and then in 1924 it was changed again, this time to Leningrad, in honor of Vladimir Lenin, the founder of Russian Communism. In 1991, following the collapse of Communist rule, the city name was changed back to St. Petersburg. Older citizens joke about being born in St. Petersburg, attending school in Petrograd, working in Leningrad, and growing old in St. Petersburg—all while living in the same place.

What independent countries were once part of Yugoslavia?

Prior to 1991, Yugoslavia was comprised of six republics: Bosnia and Herzegovina, Croatia, Macedonia, Montenegro, Serbia, and Slovenia. In 1991-92, four of the six republics—Croatia, Slovenia, Macedonia, and Bosnia and Herzegovina—declared their independence. In 2003, the remaining republics agreed to change the name from Yugoslavia to Serbia and Montenegro.

What is the Chunnel?

"Chunnel" is a nickname for the English Channel Tunnel, which connects England and France via rail under the English Channel. There are actually three separate tunnels: two for trains and a parallel service tunnel. The tunnels run for 31 miles between Coquelles, France, and Folkestone, England, at an average depth of 150 feet (46 m) below the seafloor. Work on the tunnels began in 1987, and the first trains crossed under the Channel in 1994.

Where is "Europe's Grand Canyon"?

Along the Verdon River in the Provence region of southeastern France. The Verdon has carved a deep, narrow gorge, known as the Grand Cañon du Verdon, through the limestone plateau between the town of Castellane and the artificial Lac de Ste-Croix. The gorge stretches for 13 miles (21 km) and reaches a depth of 3,170 feet (965 m). It is considered one of the natural wonders of Europe.

What is the only volcano on the European mainland that erupted in the 20th century?

Mt. Vesuvius (Vesuvio), located in southern Italy nine miles (15 km) east of Naples. It was active through much of the century, with significant eruptions in 1906, 1929, and 1944. Two thousand years ago, most Romans did not recognize Vesuvius as a volcano, and numerous farming communities thrived on the fertile land around its base. Then, in August of A.D. 79, the volcano exploded in a mighty eruption, burying the cities of Pompeii, Herculaneum, and Stabiae under cinders, ash, and mud, and killing more than 3,500 people.

How many European countries fall partially within the Arctic Circle?

The historic Henningsvaer Port in the Lofoten Islands of Norway, north of the Arctic Circle

Four: Finland, Norway, Russia, and Sweden. Technically, a fifth country could be added to this list: Iceland's mainland ends just short of the Arctic Circle, but one of the country's islands, Grimsey, straddles the line, its northern half sitting within the Circle.

Where is Waterloo, site of Napoleon's famous defeat?

Today, Waterloo is a suburb of Brussels, Belgium, although at the time of the battle—June 18, 1815—it lay 12 miles (19 km) away from the city, which was then much smaller. At Waterloo, the troops of French emperor Napoleon I were defeated by British forces under the command of the Duke of Wellington and Prussian forces led by Gebhard Blücher. The French defeat ended the Napoleonic Wars, which had begun in 1803.

Is Venice, Italy, really sinking?

Yes, although at a much slower rate than previously. The city, which dates back to the 4th century A.D., is built on 118 small islands in a lagoon at the top of the

Gondola and canal, Venice

Adriatic Sea. Its buildings sit on foundations of wooden pilings driven deep into the underlying sand, silt and clay. Originally the buildings were safely above high tide level, but over the course of 15 centuries, natural compaction of the subsoil caused the city to sink more than 30 inches (76 cm). Earlier this century, groundwater was pumped out of the subsoil to satisfy water needs on the mainland. This proved disastrous for Venice: the city quickly sank another five inches (13 cm) at a time when the sea level was rising by four inches (10 cm). The pumping was stopped, and Venice's sinking has slowed to its earlier "natural" rate. Unfortunately, the foundations of many buildings have been severely damaged by high water.

Which independent European countries are smaller than Rhode Island, the smallest U.S. state?

Seven independent European countries cover a smaller area than Rhode Island's 1,545 square miles (4,002 sq km): Vatican City, Monaco, San Marino, Liechtenstein, Malta, Andorra, and Luxembourg.

How many official languages are recognized in Switzerland?

Four: German, French, Italian, and Romansch. German is the most widely spoken language: 65% of the Swiss speak a dialect known as *Schwyzerdütsch*, or Swiss German. French is spoken by 18% of the population, Italian by 10%, and Romansch by only 1%.

Why is Ukraine called "the breadbasket of Europe"?

Ukraine's topography—flat plains, or "steppes," cover most of the country—and extremely fertile soils combine to make it one of the world's most outstanding agricultural areas. In 1994, the country produced almost 36 million tons (33 metric tons) of grain. Major crops include wheat, rye, barley, corn, potatoes, sunflower seeds, sugar beets, and cotton. Ukraine also has thriving dairy and livestock industries, as well as many food-processing plants.

If the Caspian Sea is a *sea*, then how can it be the world's largest *lake*?

Actually, it is both a sea *and* a lake. The word "sea" is used most often to designate specific regions of the oceans that are more or less surrounded by land; however, it can also apply to inland bodies of water, especially if they are large and/or salty. The Caspian Sea is both large and salty, so it is called a sea. "Lake" is a general term for inland bodies of water of substantial size. The Caspian Sea lies inland and has a surface area of 143,240 square miles (370,990 sq km), so it is also considered to be a lake. Other "sea-lakes" in the world include the Aral Sea, the Dead Sea, the Sea of Galilee, and California's Salton Sea.

Where is Transylvania?

In northwestern Romania. The region is bounded by the Carpathian Mountains in the north and east, the Transylvanian Alps in the south, and by Romania's borders with Hungary and Yugoslavia in the west. A high plateau, averaging 1,000 to 1,600 feet (300 to 500 m) in elevation, covers much of Transylvania. In Bram Stoker's 1897 novel *Dracula*, Transylvania is the home of the blood-sucking Count. Stoker based the story on local vampire legends, many of which persist today: in some parts of eastern Europe, peasants still wear garlic necklaces and hang garlic wreaths from their doors to ward off vampire spirits.

What countries contain parts of the Carpathian Mountains?

Five: the Czech Republic, Slovakia, Poland, Ukraine, and Romania. Curving for more than 900 miles (1,450 km) along the north and east sides of the Danube plain in central and eastern Europe, the Carpathians roughly form a half-circle connecting the Alps and the Balkans. The mountain system consists of two main parts: the Northern Carpathians, which include the Beskid and the Tatra ranges, and the Southern Carpathians, also called the Transylvanian Alps. The Carpathians' highest peak is Gerlachovský štít in Slovakia, which rises to 8,711 feet (2,655 m).

Where is Lapland?

In northern Scandinavia. Lapland is home to the Lapps, a nomadic people who have traditionally engaged in hunting, fishing, and reindeer-herding. When the Finns arrived in the southern part of present-day Finland 2,000 years ago, they found the Lapps already settled there. Over the years, the Lapps have been pushed north, and their territory has expanded to cover parts of northern Norway, Sweden, Finland, and northwestern Russia. Today, there are approximately 42,000 Lapps, most of whom work in a variety of farming, construction, and service fields. The Finnish government has made many efforts to protect the Lapps' language, called Sami, and culture.

What European country has a shorter coastline than any other maritime country in the world?

Harbor and coastline, Monaco

Monaco, whose Mediterranean coastline is a mere three-and-a-half miles (5.6 km) long. Another European country, Bosnia and Herzegovina, ranks second in this category: its coast on the Adriatic Sea between Croatia and Serbia and Montenegro is only 13 miles (21 km) long. In third place is Slovenia, whose Adriatic coast is 29 miles (47 km) long.

What European city is the largest city in the world north of the Arctic Circle?

Murmansk, Russia, a city of 472,900 people located on the Kola Gulf of the Barents Sea. Although Murmansk lies approximately 150 miles north of the Arctic Circle, its harbor remains ice-free throughout the year due to the moderating effect of a warm ocean current called the North Atlantic Drift. While there are thousands of cities and towns north of the Arctic Circle, there are none at all south of the Antarctic Circle at the opposite end of the world.

Questions and Answers

Africa

What is the East African Rift System?

This term refers to a series of rift valleys running through East Africa from Mozambique to the southern end of the Red Sea. These valleys are part of the Great Rift Valley, a 4,000-mile (6,430-km)-long depression that also includes the Red Sea, the Dead Sea, and the rest of the Jordan Valley. The East African Rift System marks the line along which geological forces are splitting East Africa off from the rest of the continent. Eventually, everything east of the Rift System—including all or part of present-day Mozambique, Tanzania, Rwanda, Burundi, Uganda, Kenya, Ethiopia, Djibouti, and Somalia—will be a huge island off of Africa's eastern coast. Madagascar was attached to the African mainland before similar forces split it off into an island 175 million years ago.

What is the Serengeti?

Located in northern Tanzania, east of Lake Victoria and west of Kilimanjaro, the Serengeti is a vast plain of grassland, acacia bushes, forest, and rocky outcrops. Serengeti National Park, established in 1951, covers an area of the plain about the size of Connecticut. The park is home to one of the last great concentrations of African wildlife, including antelope, buffalo, cheetahs, elephants, gazelles, giraffes, hyenas, leopards, lions, black rhinoceroses, wildebeests and zebras. Tourists from all over the world visit the park to observe the wildlife and to witness the large-scale animal migrations that occur in May and June.

Giraffes on the Serengeti

How has the Aswan High Dam affected the Nile Valley?

Before the dam was built in 1971, floodwaters inundated the Nile Valley each fall, depositing fresh, fertile silt across the valley floor. This annual replenishment of the soil helped agriculture to thrive in the valley for thousands of years. The dam ended the annual floods, and now much of the Nile's water-borne silt settles to the bottom of Lake Nassar, the enormous artificial lake behind the dam. Water evaporating from the lake's surface has increased the regional humidity, which has accelerated the decay of many of the valley's great tombs and monuments. On the positive side, the dam supplies more than 25% of Egypt's hydro-electric power, and desert irrigation projects using water from Lake Nassar have created 900,000 new acres of arable land.

What is remarkable about the delta of the Okavango River?

It is the largest inland delta in the world. The Okavango originates in the mountains of central Angola and flows 1,000 miles (1,600 km) to the northwest corner of Botswana, where it spills over the Gomare fault and fans out into a swampy delta covering 4,000 square miles (10,350 sq km). Meandering through a myriad of shallow channels, the waters of the Okavango quickly evaporate. The small amount that eventually emerges from the southeastern end of the delta represents less than 5% of the river's pre-delta flow.

What African country was previously known as Upper Volta?

Burkina Faso. The Volta River's three upper branches —the Volta Blanche (White Volta), Volta Rouge (Red Volta), and Volta Noire (Black Volta)—all originate within the country, hence the earlier name. Burkina Faso, the Mossi-dialect name adopted in 1984, translates roughly as "Country of Honest Men."

What are the most important crops grown in Africa?

Africa is the world's leading producer of cocoa beans (55% of the world total) and cassava roots (45% of the world total). It is also a major producer of grain and millet sorghum (27%), coffee (20%), peanuts (20%), palm oil (14%), tea (12%), and olive oil (12%).

Woman in sorghum fields, Bema

What object found along the banks of the Orange River in 1867 changed the course of South African history?

A 21-carat diamond. The discovery of this gem near Hopetown precipitated a huge diamond rush, and thousands of people from all over the continent and the world raced to southern Africa. The town of Kimberley, site of the famous open mine known as the Big Hole, became the diamond capital of the world. Between 1871 and 1914, more than 14 million carats of diamonds were removed from the mine, which eventually reached a depth of 4,000 feet (1,220 m) and a width of one mile (1.6 km).

What is Cabinda?

A coastal province of Angola that lies north of the Congo River and is separated from the rest of the country by a 19-mile (31-km)-wide corridor belonging to the Democratic Republic of the Congo. Most of Cabinda is covered by tropical forest. Offshore lie rich oil fields which produce one million barrels annually.

What African country was founded in 1847 by freed American slaves?

Liberia, whose name comes from the Latin word *liber*, meaning "free." It is the only country in sub-Saharan Africa that has never been ruled by a colonial power. Liberia's capital city, Monrovia, was named for James Monroe, the fifth U.S. president.

What two African countries border only a single other country?

Lesotho and The Gambia. South Africa surrounds Lesotho, and The Gambia is bordered in the north, east, and south by Senegal; to its west lies the Atlantic Ocean.

What is the Ngorongoro Crater?

Located in northern Tanzania, Ngorongoro is the crater of a volcano that has been extinct for several million years. It has a diameter of 9 miles (14.5 km) and its walls rise about 2,000 feet (610 m) above its floor. The crater supports an abundance of wildlife, including wildebeests, elephants, rhinoceroses, hippopotamuses, lions, leopards, and flamingoes. In 1956 Ngorongoro was established as a conservation area, but its ecological balance is threatened by growing numbers of tourists and the large cattle herds of the nomadic Masai people.

Q Why is it difficult to say how large Lake Chad is?

A The lake's size fluctuates dramatically throughout the year. Numerous rivers and streams flow into Lake Chad, but it has no outlet. During the summer rainy season, floodwaters swell the lake to 10,000 square miles (25,900 sq km) and occasionally to twice that size. Even at its maximum size the lake is extremely shallow; its greatest depth is only 25 feet (8 m). By the end of the following spring, evaporation has shrunk the lake by 60%, to about 4,000 square miles (10,360 sq km). In recent decades, Lake Chad's cyclical fluctuations have been greatly affected by recurring droughts, which have reduced the flow of water into the lake and accelerated evaporation. Its volume has dropped by 80% since 1970.

Q What is significant about the location of Khartoum, Sudan?

A Khartoum, the capital of Sudan, is located at the point where the White Nile and Blue Nile rivers meet to form the Nile. Capitalizing on its strategic location, the city has become Sudan's commercial center and transportation hub. It is built on a curving strip of land that resembles the trunk of an elephant: the name "Khartoum" comes from the Arabic *Ras-al-hartum*, which means "end of the elephant's trunk."

Q What is Africa's newest country?

A Eritrea, which officially became independent in 1993. An Italian colony from 1890 to 1941, Eritrea was captured by the British during World War II. In 1952, the United Nations awarded Eritrea to Ethiopia under the condition that it be ruled as a self-governing territory. Ethiopia violated this agreement by annexing Eritrea in 1962, touching off a civil war which lasted more than 30 years. Eritrea formally declared its independence in May 1993, two years after defeating Ethiopia's Marxist regime.

Q What is the Sahel?

A The Sahel (Sudan) is a semiarid region that separates the Sahara Desert from the tropical savanna and rain forests of central Africa. It stretches halfway across the continent, from Mauritania in the west to Chad in the east, in a band averaging more than 1,000 miles (1,600 km) in width. Most of the Sahel is semiarid savanna, with low grasses in the north and tall grasses in the south. Annual precipitation varies from 4 inches to 24 inches (100 to 600 mm). The 8-month dry season makes farming difficult, and the region has experienced several severe droughts in recent decades.

Woman returning from well in the Sahel

Q In what country would you find Africa's northernmost point?

A Tunisia. The northernmost point is Cape Ben Sekka, which lies just north of the continent's northernmost town, Bechater. Parts of five European countries—Greece, Italy, Malta, Portugal, and Spain—lie farther south than Cape Ben Sekka. From the tip of the cape, Africa stretches southward approximately 5,000 miles (8,000 km) to its southernmost point, Cape Agulhas in South Africa.

Q How has the Sahara Desert changed in the last 5,000 years?

A Scientists believe that 5,000 years ago the climate of the Sahara was more temperate and far less arid than it is today. Much of the region was grassland. Around 3000 B.C. global climate patterns began to shift, and the region entered an arid period which continues today. The desert currently covers 3,500,000 square miles (9,100,000 sq km), an area nearly as large as the United States, and its size is increasing. In recent decades, recurring droughts and overgrazing in the Sahel region have contributed to the Sahara's southward expansion.

Q What is the traditional mode of transportation in the Sahara Desert?

A The camel, or more specifically, the one-humped dromedary, which was domesticated at least 3,000 years ago. Dromedaries are extremely well-suited to desert conditions. They have the ability to store water in their hump, and can tolerate water losses equal to one-fourth of their body weight. Their heavy-lidded eyes and closeable nostrils offer protection in sandstorms. Today, as the Saharan road system expands, truck convoys are replacing camel caravans, although trucks require frequent refueling, often overheat in the desert sun, and grind to a halt when sand clogs their engines.

Camel eating leaves

Q Which African country can boast the greatest known deposits, variety, and output of minerals in the world?

A South Africa. It has the world's largest known deposits of chromite, gold, manganese, platinum, and vanadium. The country leads the world in production of gold, chromite, vanadium, and the platinum group metals: platinum, palladium, iridium, rhodium, and ruthenium. It is also one of the leading producers of manganese, antimony, and gem and industrial diamonds.

Q Where is the Horn of Africa?

A This term refers to the horn-shaped area of eastern Africa that juts into the Indian Ocean. Somalia and Ethiopia occupy most of the horn. The cape of Gees Gwardafuy and the city of Caluula sit at the northeastern tip of the horn, marking the entrance to the Gulf of Aden, which connects the Arabian Sea and the Red Sea.

Q What is the Valley of the Kings?

A This narrow valley, across the Nile River from the city of Luxor, contains the tombs of the pharaohs who ruled Egypt during the New Kingdom period, 1550 B.C. to 1200 B.C. The tombs are carved deep into the sandstone walls of the valley; most have five to fifteen rooms. Among the pharaohs buried in the valley are Ramses II, Ramses VI, and Seti I. Upon their death, the pharaohs were mummified and then entombed with all of the material things that they might need in the afterlife, including gold, jeweled ornaments, furniture, clothing, and food. Most of the tombs were soon looted by robbers, who removed all items of value. However, in 1922 the tomb of Tutankhamen—"King Tut"—was discovered with most of its riches untouched.

Questions and Answers

Asia

Q What natural features form the physical boundary between Europe and Asia?

A Europe and Asia share the same huge landmass, which is known as Eurasia. The imaginary line dividing this landmass into two continents runs through the Ural Mountains, the Ural River, the Caspian Sea, the Caucasus mountains, the Black Sea, the Bosporus strait, the Sea of Marmara, and the Dardanelles strait.

Q How many countries lie partially within Europe and partially within Asia?

A Four: Azerbaijan, which is traversed by the Caucasus Mountains; Kazakhstan, whose far western lands lie west of the Ural River; Russia, which is split by the Ural Mountains, and Turkey, which includes a small area on the northwestern side of the Sea of Marmara.

Q Why was the Great Wall of China built?

A To defend China against invasion by the Huns and other enemies. Defensive walls were built in China

The Great Wall winding through a hilly region in northern China

as early as the 6th century B.C. In 214 B.C., under Emperor Shih Huang-ti, the existing walls were connected to form a single continuous wall with watchtowers. This wall was extended during the Han Dynasty (202 B.C. – A.D. 220) and the Sui Dynasty (A.D. 581 – 618). Seven hundred years later the wall had mostly crumbled, and in the late 1400s, under the Ming Emperors, it was completely rebuilt. The portions of the wall that remain today are those that were constructed during this most recent period.

Q How has the Aral Sea changed in recent decades?

A It has shrunk by about 55% since 1960. The sea once covered nearly 25,000 square miles (64,720 sq km) and was the fourth-largest inland body of water in the world. Today it covers only about 11,100 square miles (28,700 sq km), and the former port city of Muynak lies 30 miles (48 km) inland. The sea's shrinkage can be blamed on cotton farming in the surrounding desert. Soviet-era efforts to establish a profitable cotton industry led to the creation of an extensive network of irrigation canals. These huge canals drain large amounts of water from the Syr Darya and Amu Darya, the only two rivers that empty into the sea.

Q What is the Ring of Fire?

A "Ring of Fire" designates the narrow band of active volcanoes encircling the Pacific Ocean basin. Of the approximately 1,500 volcanoes in the world that have been active within the last 10,000 years, more than two-thirds are part of the Ring. Over half of the Ring's active volcanoes are found in its Asian portion, which passes through Russia's Kamchatka Peninsula, the Kuril Islands, Japan, and the Philippines. The Ring of Fire's most recent major volcanic event was the 1991 eruption of Pinatubo on the Philippine island of Luzon, which prompted the evacuation of 250,000 people.

Q What part of Asia is called Indochina?

A "Indochina" refers to the southeastern Asian peninsula situated south of China and east of India. Countries located on the peninsula are Cambodia, Laos, Myanmar (Burma), Thailand, Vietnam, and the western portion of Malaysia. The eastern part of the Indochinese peninsula, including Cambodia, Laos, and Vietnam, was formerly known as French Indochina because of France's strong colonial presence there.

Q The Khyber Pass links which two countries?

A Afghanistan and Pakistan. Approximately 33 miles (53 km) long and reaching a maximum elevation of about 3,500 feet (1,067 m), the pass cuts through the Safed Koh mountains just south of the Kabul River, connecting the high plateau of Afghanistan with the Indus Valley. It has been used for centuries as a caravan route and as an invasion route into India. Today it is also traversed by a paved highway and, in Pakistan, by a railroad. In the 1980s several million refugees fleeing Afghanistan's civil war crossed into Pakistan via the pass.

Q What Persian Gulf country is a federation of seven Arab sheikdoms?

A United Arab Emirates, formed in 1971 through the unification of the sheikdoms of Abu Dhabi, Ajman, Dubai, Fujeirah, Sharjah, and Umm al-Qawain. Ras al-Khaimah joined the federation in 1972. Underdeveloped a few decades ago, the U.A.E. has been transformed by oil wealth into a modern and affluent country.

Q What Asian country contains, or is bordered by, six of the world's ten highest mountains?

A Nepal. Within Nepal or along its borders with China and India are found the following peaks: Mt. Everest (highest in the world), Kanchenjunga (3rd), Makalu (4th), Dhawalāgiri (5th), Annapurna (7th), and Xixabangma Feng (9th). Nanda Devi (10th) lies only 50 miles (80 km) northwest of Nepal's western border.

Q Where is the Empty Quarter?

A This hostile desert is found in the southern third of the Arabian Peninsula. Called *Ar Rub' Al-Khālī*, or "the Empty Quarter" in Arabic, it covers 250,000 square miles (647,000 sq km) and is the world's largest continuous sand body. Few people live in the Empty Quarter, and much of the region has never been explored.

Q What Asian volcanic eruption has been called the loudest natural explosion in recorded history?

A The 1883 eruption of Krakatau (Krakatoa), an island volcano between Sumatra and Java. Krakatau exploded three times on August 26 and 27, 1883, shooting tremendous amounts of gas and ash 50 miles (80 km) into the atmosphere. The explosions were so violent that they were heard nearly 3,000 miles (4,653 km) away on Rodrigues Island in the western Indian Ocean. Krakatau collapsed into itself, and when the explosions were over most of the island was submerged under 900 feet of water. Tsunamis up to 130 feet (40 m) high slammed the coasts of Sumatra and Java, washing away hundreds of villages and killing more than 36,000 people.

At their closest point, how far apart are Asia and North America?

At the narrowest point of the Bering Strait, Asia and North America are separated by only 56 miles (90 km). Russia's Big Diomede (Ratmanov) Island and the United States' Little Diomede Island, which lie in the middle of the strait, are only 2.5 miles (4 km) apart.

Where is the Fertile Crescent?

This term refers to a crescent-shaped area of fertile land in the Middle East which begins in the south with Egypt's Nile Valley, runs north along the eastern coast of the Mediterranean Sea, then turns southeast through Mesopotamia, the land between the Tigris and Euphrates rivers, and ends at the head of the Persian Gulf. The Fertile Crescent was the birthplace of some of the world's oldest civilizations, including the Sumerians, Babylonians, and Assyrians.

What country was Pakistan part of before it became independent?

India. Conflicts between Hindus and Muslims in British India led to the creation of Pakistan as a separate Muslim state in 1947. Originally, Pakistan included the two main centers of Muslim population, which lay in northwest and east India. The two areas, West Pakistan and East Pakistan, were separated by 1,000 miles (1,600 km). In 1971, East Pakistan declared its independence and changed its name to Bangladesh. The name "Pakistan" comes from the Urdu words *pakh*, meaning "pure," and *stan*, meaning "land."

What was Sri Lanka called before 1972?

Ceylon, which is the name the British had given to the island when they claimed it in 1796. The island became independent in 1948, and in 1972 was renamed Sri Lanka, which in the Sinhala language means "Resplendent Land."

The Taj Mahal

What is India's most famous tomb?

The Taj Mahal, located in the city of Agra. Often described as one of the world's most beautiful buildings, it was built by the Mogul emperor Shah Jahan to honor the memory of his wife, Mumtaz-i-Mahal. Construction began in 1631 and was completed in 1648.

What two seas are linked by the Suez Canal?

The Red Sea and the Mediterranean Sea. Before the 101-mile (163-km) canal was built in the mid-1800s, ships traveling between Europe and the Far East had to sail all the way around the southern tip of Africa. Depending on the origin and destination of the ship, the canal could shorten its trip dramatically. For example, a ship sailing from London to Bombay would have to travel almost 11,000 miles (17,700 km) around the African continent. Using the Suez Canal, the trip could be shortened to 6,300 miles (10,140 km), a distance reduction of over 40%.

Which independent Asian country has the highest population density?

The tiny republic of Singapore, with 17,814 people per square mile (6,879 per sq km). Singapore's 4,375,000 people occupy an island measuring only 26 miles east-to-west and 14 miles north-to-south. Mongolia has the lowest density: 4.4 people per square mile (1.7 per sq km). Mongolia's land area is 2,458 times larger than Singapore's, but it holds only 2,675,000 people.

How long is Japan, from north to south?

The islands of Japan stretch approximately 1,900 miles (3,060 km) from Hokkaido in the north to the Sakishima Archipelago in the south. This is approximately equal to the distance between New York City and Denver.

Landscape on Hokkaido, Japan's northernmost island

What region is known as the "Roof of the World"?

Tibet, the high plateau region which lies north of the Himalayas. Covering 471,000 square miles (1,220,000 sq km), and with an average elevation of 15,000 feet (4,600 m), the Tibetan Plateau is the largest and highest plateau in the world. Much of Tibet is uninhabited; the region's fewer than two million people are concentrated in the valleys of the Brahmaputra (Yarlung) River and its tributaries.

Which Asian country leads the world in number of earthquake-related deaths since 1900?

China, where 48 deadly earthquakes have killed an estimated 967,420 people since 1900. During the same period Iran has lost 147,293 people in 57 earthquakes, and in Japan 32 earthquakes have killed 123,462 people.

What cities mark the endpoints of the Trans-Siberian Railway?

The longest railway line in the world, the Trans-Siberian Railway stretches 5,764 miles (9,297 km) between Moscow in the west and the Pacific Coast port city of Nakhodka (near Vladivostok) in the east. The eight-day journey between the two cities includes stops in 92 Russian cities and towns.

How many people live on the Indonesian island of Java?

Java is home to approximately 118 million people. It is the world's most populous island, although it is only the 13th-largest in area. By contrast, the island of Cuba is four-fifths the size of Java, but its population is only one-eleventh as large.

What was the name of Ho Chi Minh City, Vietnam, prior to 1975?

Saigon. When the city fell to North Vietnamese forces in 1975, it was renamed for Ho Chi Minh, founder of the Indochina Communist Party and president of North Vietnam from 1945 until his death in 1969.

Questions and Answers

Oceania (including Australia and New Zealand)

Q What is Oceania?

A The name "Oceania" refers to the scattered islands of a vast area of the Pacific Ocean, from Palau in the west to Easter Island in the east, and from the Midway Islands in the north to New Zealand in the south. The three main island groups of Oceania are Melanesia, Micronesia, and Polynesia. The continent of Australia and the islands of New Zealand are sometimes considered part of Oceania.

Q What is the Outback?

A This nickname refers to Australia's vast, largely uninhabited interior. Over the years, the Outback's harsh beauty and its remoteness from the rest of Australia have made it popular with adventurous explorers and travelers. Through depictions in literature, art, and film, the region has become an integral part of Australia's identity. However, it is difficult to characterize the Outback, for its boundaries are undefined and its landscape varies from hot deserts to lush wilderness.

Eucalyptus tree in the Outback, Northern Territory

Q Why is Australia sometimes referred to as "the Land Down Under"?

A This nickname originated with the British, who began colonizing Australia in the late 1700s. Because of its extreme southern location in relation to Britain, Australia was considered "Down." The "Under" part of the phrase refers to the continent's position "under" the Eurasian landmass. But while people from the Northern Hemisphere think of Australia as "Down Under," Australians do not.

Q How much of Australia is arid or semiarid?

A More than two-thirds of the continent is considered to be arid or semiarid. The arid areas comprise several large deserts, including the Great Victoria Desert, the Gibson Desert, the Great Sandy Desert, and the Simpson Desert.

Q What distinction does Wellington, New Zealand have among national capitals of independent countries?

A Located at 41°18' south latitude, Wellington is the world's southernmost national capital. In second place is Canberra, Australia, which lies at 35°17' south latitude.

Q What is unusual about how the island of Nauru was formed?

A Nauru, the world's smallest republic, began as a coral atoll. Over the millennia, accumulated bird droppings filled in the central lagoon and created an 8-square-mile (21-sq-km) island whose highest point rises 210 feet (64 m) above sea level. The droppings are a rich source of phosphate, which is used in making fertilizers. Phosphate mining has long been Nauru's economic mainstay, but the resource will soon be exhausted.

Q What are the principal islands of New Zealand?

A Two islands, North Island and South Island, account for more than 98% of New Zealand's total land area. The country also includes Stewart Island, the Chatham Islands, the Antipodes Islands, the Auckland Islands, and hundreds of tiny islets.

Q What is Ayers Rock?

A Ayers Rock, now called by its Aboriginal name, Uluru, is a huge, red, oval-shaped rock outcropping that rises 2,831 feet (863 m) above the plains of central Australia. One of the largest monoliths in the world, Uluru is actually the summit of a massive sandstone hill, most of which is hidden underground. Aborigines consider Uluru sacred and incorporate numerous places around it into their ceremonial life.

Q What is the ratio of sheep to humans in Australia and New Zealand?

A Because both countries are major wool, mutton, and lamb producers, Australia and New Zealand each have a high ratio of sheep to humans. In Australia, there are 132 million sheep, or seven sheep for each human in the country. In New Zealand, the ratio is even greater: the country's 50 million sheep translate into 14 sheep for each human.

Q How many of Oceania's countries have become independent since 1975?

A Eight. Papua New Guinea became independent in 1975, the Solomon Islands and Tuvalu in 1978, Kiribati in 1979, Vanuatu in 1980, the Marshall Islands and the Federated States of Micronesia in 1986, and Palau (Belau) in 1994.

Q What is the Great Barrier Reef?

A This vast coral reef system is the longest in the world, stretching 1,250 miles (2,000 km) along Australia's northeast coast. It is composed of reefs, shoals, and hundreds of islands. Popular with divers, the Great Barrier Reef is home to a myriad of aquatic creatures.

Q On what island is Robert Louis Stevenson buried?

A Stevenson, author of *Dr. Jekyll and Mr. Hyde*, *Kidnapped*, and *Treasure Island*, is buried on the island of Upolu in Western Samoa. Born in Edinburgh, Scotland in 1850, Stevenson sailed to the South Pacific in 1888 and settled permanently in Samoa in 1890. He died there in 1894.

Q Where does Sydney rank among Australia's most populous cities?

A Measured by actual city population, Sydney is not even ranked in the top hundred: only 13,501 people live within its tiny city limits. Brisbane is Australia's largest city, with 751,115 people. The Sydney metropolitan area, however, is by far the largest in Australia, with 3,538,749 people.

Harbor and skyline of Sydney

Q What is Australia's only island state?

A Tasmania, which lies about 150 miles (240 km) south of the Australian mainland. Measuring 26,200 square miles (67,800 sq km) in area, it is the smallest of Australia's six states and accounts for less than 1% of the country's area. Originally part of New South Wales, Tasmania became a separate colony in 1825 and a state in 1901. Hobart, its capital, is home to 45% of the state's population.

Countries and Flags

This 12-page section presents basic information about each of the world's countries, along with an illustration of each country's flag. A total of 198 countries are listed: the world's 191 fully independent countries, and 7 internally independent countries which are under the protection of other countries in matters of defense and foreign affairs. Colonies and other dependent political entities are not listed. The design of the East Timor flag has not been finalized.

The categories of information provided for each country are as follows.

Flag: In many countries two or more versions of the national flag exist. For example, there is often a "civil" version which the average person flies, and a "state" version which is flown only at government buildings and government functions. A common difference between the two is the inclusion of a coat of arms on the state version.

Country name: The short form of the English translation of the official country name.

Official name: The long form of the English translation of the official country name.

Population: The population figures listed are 2002 estimates based on U.S. census bureau figures and other available information.

Area: Figures provided represent total land area and all inland water. They are based on official data or U.N. data.

Population density: The number of people per square mile and square kilometer, calculated by dividing the country's population figure by its area figure.

Capital: The city that serves as the official seat of government. Population figures follow the capital name. These figures are based upon the latest official data.

AFGHANISTAN
Official Name: Islamic State of Afghanistan
Population: 27,280,000
Area: 251,826 sq mi (652,225 sq km)
Density: 108/sq mi (42/sq km)
Capital: Kābul, 1,424,400

ALGERIA
Official Name: Democratic and Popular Republic of Algeria
Population: 32,005,000
Area: 919,595 sq mi (2,381,741 sq km)
Density: 35/sq mi (13/sq km)
Capital: Algiers (El Djazaïr),1,507,241

ANGUILLA
Official Name: Anguilla
Population: 12,000
Area: 35 sq mi (91 sq km)
Density: 342/sq mi (132/sq km)
Capital: The Valley, 1,462

ALBANIA
Official Name: Republic of Albania
Population: 3,525,000
Area: 11,100 sq mi (28,748 sq km)
Density: 318/sq mi (123/sq km)
Capital: Tiranë, 243,000

ANDORRA
Official Name: Principality of Andorra
Population: 68,000
Area: 175 sq mi (453 sq km)
Density: 389/sq mi (150/sq km)
Capital: Andorra, 20,437

ANTIGUA AND BARBUDA
Official Name: Antigua and Barbuda
Population: 67,000
Area: 171 sq mi (442 sq km)
Density: 393/sq mi (152/sq km)
Capital: St. John's, 24,359

ANGOLA
Official Name: Republic of Angola
Population: 10,480,000
Area: 481,354 sq mi (1,246,700 sq km)
Density: 22/sq mi (8.4/sq km)
Capital: Luanda, 1,459,900

Countries and Flags *continued*

ARGENTINA
Official Name: Argentine Republic
Population: 37,600,000
Area: 1,073,519 sq mi (2,780,400 sq km)
Pop. Density: 35/sq mi (14/sq km)
Capital: Buenos Aires (de facto), 2,960,976,
and Viedma (future), 40,452

ARMENIA
Official Name: Republic of Armenia
Population: 3,335,000
Area: 11,506 sq mi (29,800 sq km)
Pop. Density: 290/sq mi (112/sq km)
Capital: Yerevan, 1,199,000

AUSTRALIA
Official Name: Commonwealth of Australia
Population: 19,455,000
Area: 2,969,910 sq mi (7,692,030 sq km)
Pop. Density: 6.6/sq mi (2.5/sq km)
Capital: Canberra, 298,847

AUSTRIA
Official Name: Republic of Austria
Population: 8,160,000
Area: 32,378 sq mi (83,859 sq km)
Pop. Density: 252/sq mi (97/sq km)
Capital: Vienna (Wien), 1,539,848

AZERBAIJAN
Official Name: Azerbaijani Republic
Population: 7,785,000
Area: 33,437 sq mi (86,600 sq km)
Pop. Density: 233/sq mi (90/sq km)
Capital: Baku (Bakı), 1,080,500

BAHAMAS
Official Name: Commonwealth of the
Bahamas
Population: 300,000
Area: 5,382 sq mi (13,939 sq km)
Pop. Density: 56/sq mi (22/sq km)
Capital: Nassau, 141,000

BAHRAIN
Official Name: State of Bahrain
Population: 650,000
Area: 267 sq mi (691 sq km)
Pop. Density: 2,436/sq mi (941/sq km)
Capital: Al Manāmah, 127,578

BANGLADESH
Official Name: People's Republic of
Bangladesh
Population: 132,315,000
Area: 55,598 sq mi (143,998 sq km)
Pop. Density: 2,380/sq mi (919/sq km)
Capital: Dhaka (Dacca), 3,637,892

BARBADOS
Official Name: Barbados
Population: 275,000
Area: 166 sq mi (430 sq km)
Pop. Density: 1,657/sq mi (640/sq km)
Capital: Bridgetown, 5,928

BELARUS
Official Name: Republic of Belarus
Population: 10,340,000
Area: 80,155 sq mi (207,600 sq km)
Pop. Density: 129/sq mi (50/sq km)
Capital: Minsk, 1,661,000

BELGIUM
Official Name: Kingdom of Belgium
Population: 10,265,000
Area: 11,787 sq mi (30,528 sq km)
Pop. Density: 871/sq mi (336/sq km)
Capital: Brussels (Bruxelles), 136,424

BELIZE
Official Name: Belize
Population: 260,000
Area: 8,866 sq mi (22,963 sq km)
Pop. Density: 29/sq mi (11/sq km)
Capital: Belmopan, 8,130

BENIN
Official Name: Republic of Benin
Population: 6,690,000
Area: 43,475 sq mi (112,600 sq km)
Pop. Density: 154/sq mi (59/sq km)
Capital: Porto-Novo (designated), 179,138,
and Cotonou (de facto), 536,827

BHUTAN
Official Name: Kingdom of Bhutan
Population: 2,070,000
Area: 17,954 sq mi (46,500 sq km)
Pop. Density: 115/sq mi (45/sq km)
Capital: Thimphu, 12,000

BOLIVIA
Official Name: Republic of Bolivia
Population: 8,375,000
Area: 424,165 sq mi (1,098,581 sq km)
Pop. Density: 20/sq mi (7.6/sq km)
Capital: La Paz (seat of government),
792,611, and Sucre (legal capital), 194,888

BOSNIA AND HERZEGOVINA
Official Name: Republic of Bosnia and
Herzegovina
Population: 3,950,000
Area: 19,741 sq mi (51,129 sq km)
Pop. Density: 200/sq mi (77/sq km)
Capital: Sarajevo, 367,703

BOTSWANA
Official Name: Republic of Botswana
Population: 1,590,000
Area: 224,712 sq mi (582,000 sq km)
Pop. Density: 7.1/sq mi (2.7/sq km)
Capital: Gaborone, 133,468

BRAZIL
Official Name: Federative Republic of Brazil
Population: 175,260,000
Area: 3,300,172 sq mi (8,547,404 sq km)
Pop. Density: 53/sq mi (21/sq km)
Capital: Brasília, 1,947,133

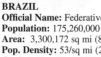

BRUNEI
Official Name: Negara Brunei Darussalam
Population: 345,000
Area: 2,226 sq mi (5,765 sq km)
Pop. Density: 155/sq mi (60/sq km)
Capital: Bandar Seri Begawan, 45,867

BULGARIA
Official Name: Republic of Bulgaria
Population: 7,665,000
Area: 42,855 sq mi (110,994 sq km)
Pop. Density: 179/sq mi (69/sq km)
Capital: Sofia (Sofiya), 1,190,126

BURKINA FASO
Official Name: Burkina Faso
Population: 12,435,000
Area: 105,869 sq mi (274,200 sq km)
Pop. Density: 117/sq mi (45/sq km)
Capital: Ouagadougou, 441,514

BURUNDI
Official Name: Republic of Burundi
Population: 6,300,000
Area: 10,745 sq mi (27,830 sq km)
Pop. Density: 586/sq mi (226/sq km)
Capital: Bujumbura, 226,628

CAMBODIA
Official Name: Kingdom of Cambodia
Population: 12,630,000
Area: 69,898 sq mi (181,035 sq km)
Pop. Density: 181/sq mi (70/sq km)
Capital: Phnom Penh (Phnum Pénh),
  620,000

CAMEROON
Official Name: Republic of Cameroon
Population: 15,995,000
Area: 183,568 sq mi (475,440 sq km)
Pop. Density: 87/sq mi (34/sq km)
Capital: Yaoundé, 620,000

CANADA
Official Name: Canada
Population: 31,750,000
Area: 3,855,103 sq mi (9,984,670 sq km)
Pop. Density: 8.2/sq mi (3.2/sq km)
Capital: Ottawa, 323,340

CAPE VERDE
Official Name: Republic of Cape Verde
Population: 405,000
Area: 1,557 sq mi (4,033 sq km)
Pop. Density: 260/sq mi (100/sq km)
Capital: Praia, 61,644

CENTRAL AFRICAN REPUBLIC
Official Name: Central African Republic
Population: 3,610,000
Area: 240,536 sq mi (622,984 sq km)
Pop. Density: 15/sq mi (5.8/sq km)
Capital: Bangui, 451,690

CHAD
Official Name: Republic of Chad
Population: 8,850,000
Area: 495,755 sq mi (1,284,000 sq km)
Pop. Density: 18/sq mi (6.9/sq km)
Capital: N'Djamena, 546,572

CHILE
Official Name: Republic of Chile
Population: 15,415,000
Area: 292,135 sq mi (756,626 sq km)
Pop. Density: 53/sq mi (20/sq km)
Capital: Santiago, 4,295,593

CHINA
Official Name: People's Republic of China
Population: 1,278,720,000
Area: 3,690,045 sq mi (9,557,172 sq km)
Pop. Density: 347/sq mi (134/sq km)
Capital: Beijing (Peking), 6,690,000

COLOMBIA
Official Name: Republic of Colombia
Population: 40,680,000
Area: 440,831 sq mi (1,141,748 sq km)
Pop. Density: 92/sq mi (36/sq km)
Capital: Bogotá, 4,931,796

COMOROS
Official Name: Federal Islamic Republic of the
  Comoros
Population: 605,000
Area: 863 sq mi (2,235 sq km)
Pop. Density: 701/sq mi (271/sq km)
Capital: Moroni, 23,432

CONGO
Official Name: Republic of the Congo
Population: 2,925,000
Area: 132,047 sq mi (342,000 sq km)
Pop. Density: 22/sq mi (8.6/sq km)
Capital: Brazzaville, 693,712

**CONGO, DEMOCRATIC REPUBLIC OF
  THE**
Official Name: Democratic Republic
  of the Congo
Population: 54,455,000
Area: 905,446 sq mi (2,345,095 sq km)
Pop. Density: 60/sq mi (23/sq km)
Capital: Kinshasa, 3,000,000

COSTA RICA
Official Name: Republic of Costa Rica
Population: 3,805,000
Area: 19,730 sq mi (51,100 sq km)
Pop. Density: 193/sq mi (74/sq km)
Capital: San José, 309,672

Countries and Flags *continued*

COTE D'IVOIRE
Official Name: Republic of Cote d'Ivoire
Population: 16,600,000
Area: 124,518 sq mi (322,500 sq km)
Pop. Density: 133/sq mi (51/sq km)
Capital: Abidjan (de facto), 1,929,079, and
Yamoussoukro (designated), 106,786

CROATIA
Official Name: Republic of Croatia
Population: 4,365,000
Area: 21,829 sq mi (56,538 sq km)
Pop. Density: 200/sq mi (77/sq km)
Capital: Zagreb, 867,865

CUBA
Official Name: Republic of Cuba
Population: 11,205,000
Area: 42,804 sq mi (110,861 sq km)
Pop. Density: 262/sq mi (101/sq km)
Capital: Havana (La Habana), 2,189,716

CYPRUS
Official Name: Republic of Cyprus
Population: 625,000
Area: 2,277 sq mi (5,896 sq km)
Pop. Density: 275/sq mi (106/sq km)
Capital: Nicosia (Levkosía), 47,036

CYPRUS, NORTH
Official Name: Turkish Republic of
Northern Cyprus
Population: 140,000
Area: 1,295 sq mi (3,355 sq km)
Pop. Density: 108/sq mi (42/sq km)
Capital: Nicosia (Lefkoşa), 37,400

CZECH REPUBLIC
Official Name: Czech Republic
Population: 10,260,000
Area: 30,450 sq mi (78,866 sq km)
Pop. Density: 337/sq mi (130/sq km)
Capital: Prague (Praha), 1,214,174

DENMARK
Official Name: Kingdom of Denmark
Population: 5,360,000
Area: 16,639 sq mi (43,094 sq km)
Pop. Density: 322 sq mi (124/sq km)
Capital: Copenhagen (København), 487,969

DJIBOUTI
Official Name: Republic of Djibouti
Population: 465,000
Area: 8,958 sq mi (23,200 sq km)
Pop. Density: 52/sq mi (20/sq km)
Capital: Djibouti, 329,337

DOMINICA
Official Name: Commonwealth of Dominica
Population: 70,000
Area: 305 sq mi (790 sq km)
Pop. Density: 230/sq mi (89/sq km)
Capital: Roseau, 9,348

DOMINICAN REPUBLIC
Official Name: Dominican Republic
Population: 8,650,000
Area: 18,704 sq mi (48,442 sq km)
Pop. Density: 462/sq mi (179/sq km)
Capital: Santo Domingo, 1,609,966

ECUADOR
Official Name: Republic of Ecuador
Population: 13,315,000
Area: 105,037 sq mi (272,045 sq km)
Pop. Density: 127/sq mi (49/sq km)
Capital: Quito, 1,100,847

EGYPT
Official Name: Arab Republic of Egypt
Population: 70,125,000
Area: 386,662 sq mi (1,001,449 sq km)
Pop. Density: 181/sq mi (70/sq km)
Capital: Cairo (Al Qāhirah), 6,801,695

EL SALVADOR
Official Name: Republic of El Salvador
Population: 6,295,000
Area: 8,124 sq mi (21,041 sq km)
Pop. Density: 775/sq mi (299/sq km)
Capital: San Salvador, 415,346

EQUATORIAL GUINEA
Official Name: Republic of Equatorial Guinea
Population: 490,000
Area: 10,831 sq mi (28,051 sq km)
Pop. Density: 45/sq mi (17/sq km)
Capital: Malabo, 31,630

ERITREA
Official Name: State of Eritrea
Population: 4,380,000
Area: 36,170 sq mi (93,679 sq km)
Pop. Density: 121/sq mi (47/sq km)
Capital: Asmera, 358,100

ESTONIA
Official Name: Republic of Estonia
Population: 1,420,000
Area: 17,413 sq mi (45,100 sq km)
Pop. Density: 82/sq mi (31/sq km)
Capital: Tallinn, 403,981

ETHIOPIA
Official Name: Ethiopia
Population: 66,780,000
Area: 446,953 sq mi (1,157,603 sq km)
Pop. Density: 149/sq mi (58/sq km)
Capital: Addis Ababa (Adis Abeba), 2,084,588

FIJI
Official Name: Republic of Fiji
Population: 850,000
Area: 7,056 sq mi (18,274 sq km)
Pop. Density: 120/sq mi (47/sq km)
Capital: Suva, 77,366

FINLAND
Official Name: Republic of Finland
Population: 5,180,000
Area: 130,559 sq mi (338,145 sq km)
Pop. Density: 40/sq mi (15/sq km)
Capital: Helsinki (Helsingfors), 512,686

FRANCE
Official Name: French Republic
Population: 59,660,000
Area: 208,482 sq mi (539,965 sq km)
Pop. Density: 286/sq mi (110/sq km)
Capital: Paris, 2,147,857

GABON
Official Name: Gabonese Republic
Population: 1,225,000
Area: 103,347 sq mi (267,667 sq km)
Pop. Density: 12/sq mi (4.6/sq km)
Capital: Libreville, 337,700

THE GAMBIA
Official Name: Republic of the Gambia
Population: 1,435,000
Area: 4,127 sq mi (10,689 sq km)
Pop. Density: 348/sq mi (134/sq km)
Capital: Banjul, 42,407

GEORGIA
Official Name: Republic of Georgia
Population: 4,975,000
Area: 26,911 sq mi (69,700 sq km)
Pop. Density: 185/sq mi (71/sq km)
Capital: Tbilisi, 1,279,000

GERMANY
Official Name: Federal Republic of Germany
Population: 83,145,000
Area: 137,822 sq mi (356,955 sq km)
Pop. Density: 603/sq mi (233/sq km)
Capital: Berlin, 3,425,759

GHANA
Official Name: Republic of Ghana
Population: 20,070,000
Area: 92,098 sq mi (238,533 sq km)
Pop. Density: 218/sq mi (84/sq km)
Capital: Accra, 949,113

GREECE
Official Name: Hellenic Republic
Population: 10,635,000
Area: 50,949 sq mi (131,957 sq km)
Pop. Density: 209/sq mi (81/sq km)
Capital: Athens (Athínai), 772,072

GREENLAND
Official Name: Greenland
Population: 56,000
Area: 840,004 sq mi (2,175,600 sq km)
Pop. Density: 0.07/sq mi (0.03/sq km)
Capital: Godthåb (Nuuk), 13,445

GRENADA
Official Name: Grenada
Population: 89,000
Area: 133 sq mi (344 sq km)
Pop. Density: 670/sq mi (259/sq km)
Capital: St. George's, 4,439

GUATEMALA
Official Name: Republic of Guatemala
Population: 13,145,000
Area: 42,042 sq mi (108,889 sq km)
Pop. Density: 313/sq mi (121/sq km)
Capital: Guatemala, 823,301

GUINEA
Official Name: Republic of Guinea
Population: 7,690,000
Area: 94,926 sq mi (245,857 sq km)
Pop. Density: 81/sq mi (31/sq km)
Capital: Conakry, 950,000

GUINEA-BISSAU
Official Name: Republic of Guinea-Bissau
Population: 1,330,000
Area: 13,948 sq mi (36,125 sq km)
Pop. Density: 95/sq mi (37/sq km)
Capital: Bissau, 125,000

GUYANA
Official Name: Co-operative Republic of Guyana
Population: 695,000
Area: 83,000 sq mi (214,969 sq km)
Pop. Density: 8.4/sq mi (3.2/sq km)
Capital: Georgetown, 78,500

HAITI
Official Name: Republic of Haiti
Population: 7,015,000
Area: 10,714 sq mi (27,750 sq km)
Pop. Density: 655/sq mi (253/sq km)
Capital: Port-au-Prince, 846,247

HONDURAS
Official Name: Republic of Honduras
Population: 6,485,000
Area: 43,277 sq mi (112,088 sq km)
Pop. Density: 150/sq mi (58/sq km)
Capital: Tegucigalpa, 576,661

HUNGARY
Official Name: Republic of Hungary
Population: 10,090,000
Area: 35,919 sq mi (93,030 sq km)
Pop. Density: 281/sq mi (108/sq km)
Capital: Budapest, 1,906,798

ICELAND
Official Name: Republic of Iceland
Population: 280,000
Area: 39,769 sq mi (103,000 sq km)
Pop. Density: 7.0/sq mi (3.0/sq km)
Capital: Reykjavik, 100,850

Countries and Flags *continued*

INDIA
Official Name: Republic of India
Population: 1,037,955,000
Area: 1,222,559 sq mi (3,166,414 sq km)
Pop. Density: 849/sq mi (328/sq km)
Capital: New Delhi, 294,783

INDONESIA
Official Name: Republic of Indonesia
Population: 230,260,000
Area: 735,310 sq mi (1,904,443 sq km)
Pop. Density: 313/sq mi (121/sq km)
Capital: Jakarta, 8,227,746

IRAN
Official Name: Islamic Republic of Iran
Population: 66,365,000
Area: 630,578 sq mi (1,633,189 sq km)
Pop. Density: 105/sq mi (41/sq km)
Capital: Tehrān, 6,758,845

IRAQ
Official Name: Republic of Iraq
Population: 23,665,000
Area: 169,235 sq mi (438,317 sq km)
Pop. Density: 140/sq mi (54/sq km)
Capital: Baghdād, 3,841,268

IRELAND
Official Name: Ireland
Population: 3,860,000
Area: 27,133 sq mi (70,273 sq km)
Pop. Density: 142/sq mi (55/sq km)
Capital: Dublin (Baile Átha Cliath), 481,854

ISRAEL
Official Name: State of Israel
Population: 5,985,000
Area: 8,019 sq mi (20,770 sq km)
Pop. Density: 746/sq mi (288/sq km)
Capital: Jerusalem (Yerushalayim), 633,700

ITALY
Official Name: Italian Republic
Population: 57,700,000
Area: 116,342 sq mi (301,323 sq km)
Pop. Density: 496/sq mi (191/sq km)
Capital: Rome (Roma), 2,649,765

JAMAICA
Official Name: Jamaica
Population: 2,670,000
Area: 4,244 sq mi (10,991 sq km)
Pop. Density: 629/sq mi (243/sq km)
Capital: Kingston, 516,500

JAPAN
Official Name: Japan
Population: 126,880,000
Area: 145,850 sq mi (377,750 sq km)
Pop. Density: 870/sq mi (336/sq km)
Capital: Tōkyō, 7,967,614

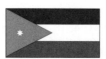

JORDAN
Official Name: Hashemite Kingdom of
Jordan
Population: 5,230,000
Area: 35,135 sq mi (91,000 sq km)
Pop. Density: 149/sq mi (57/sq km)
Capital: 'Ammān, 963,490

KAZAKHSTAN
Official Name: Republic of Kazakhstan
Population: 16,735,000
Area: 1,049,156 sq mi (2,717,300 sq km)
Pop. Density: 16/sq mi (6.2/sq km)
Capital: Astana, 286,000

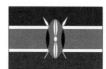

KENYA
Official Name: Republic of Kenya
Population: 30,960,000
Area: 224,961 sq mi (582,646 sq km)
Pop. Density: 138/sq mi (53/sq km)
Capital: Nairobi, 2,143,254

KIRIBATI
Official Name: Republic of Kiribati
Population: 95,000
Area: 313 sq mi (811 sq km)
Pop. Density: 303/sq mi (117/sq km)
Capital: Bairiki, 2,226

KOREA, NORTH
Official Name: Democratic People's Republic
of Korea
Population: 22,100,000
Area: 46,540 sq mi (120,538 sq km)
Pop. Density: 475/sq mi (183/sq km)
Capital: P'yŏngyang, 2,355,000

KOREA, SOUTH
Official Name: Republic of Korea
Population: 48,120,000
Area: 38,230 sq mi (99,016 sq km)
Pop. Density: 1,259/sq mi (486/sq km)
Capital: Seoul (Sŏul), 10,627,790

KUWAIT
Official Name: State of Kuwait
Population: 2,075,000
Area: 6,880 sq mi (17,818 sq km)
Pop. Density: 302/sq mi (116/sq km)
Capital: Kuwait (Al Kuwayt), 28,859

KYRGYZSTAN
Official Name: Kyrgyz Republic
Population: 4,785,000
Area: 76,641 sq mi (198,500 sq km)
Pop. Density: 62/sq mi (24/sq km)
Capital: Bishkek, 631,300

LAOS
Official Name: Lao People's Democratic
Republic
Population: 5,705,000
Area: 91,429 sq mi (236,800 sq km)
Pop. Density: 62/sq mi (24/sq km)
Capital: Viangchan (Vientiane), 464,000

LATVIA
Official Name: Republic of Latvia
Population: 2,375,000
Area: 24,595 sq mi (63,700 sq km)
Pop. Density: 97/sq mi (37/sq km)
Capital: Rīga, 874,200

LEBANON
Official Name: Republic of Lebanon
Population: 3,655,000
Area: 4,016 sq mi (10,400 sq km)
Pop. Density: 910/sq mi (351/sq km)
Capital: Beirut (Bayrūt), 509,000

LESOTHO
Official Name: Kingdom of Lesotho
Population: 2,195,000
Area: 11,720 sq mi (30,355 sq km)
Pop. Density: 187/sq mi (72/sq km)
Capital: Maseru, 137,837

LIBERIA
Official Name: Republic of Liberia
Population: 3,255,000
Area: 38,250 sq mi (99,067 sq km)
Pop. Density: 85/sq mi (33/sq km)
Capital: Monrovia, 465,000

LIBYA
Official Name: Socialist People's Libyan
Arab Jamahiriya
Population: 5,305,000
Area: 679,362 sq mi (1,759,540 sq km)
Pop. Density: 7.8/sq mi (3.0/sq km)
Capital: Tripoli (Tarābulus), 591,062

LIECHTENSTEIN
Official Name: Principality of Liechtenstein
Population: 33,000
Area: 62 sq mi (160 sq km)
Pop. Density: 534/sq mi (206/sq km)
Capital: Vaduz, 5,106

LITHUANIA
Official Name: Republic of Lithuania
Population: 3,605,000
Area: 25,213 sq mi (65,300 sq km)
Pop. Density: 143/sq mi (55/sq km)
Capital: Vilnius, 578,639

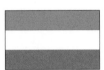

LUXEMBOURG
Official Name: Grand Duchy of Luxembourg
Population: 445,000
Area: 999 sq mi (2,586 sq km)
Pop. Density: 446/sq mi (172/sq km)
Capital: Luxembourg, 81,800

MACEDONIA
Official Name: Republic of Macedonia
Population: 2,050,000
Area: 9,928 sq mi (25,713 sq km)
Pop. Density: 206/sq mi (80/sq km)
Capital: Skopje, 440,577

MADAGASCAR
Official Name: Republic of Madagascar
Population: 16,225,000
Area: 226,658 sq mi (587,041 sq km)
Pop. Density: 72/sq mi (28/sq km)
Capital: Antananarivo, 1,250,000

MALAWI
Official Name: Republic of Malawi
Population: 10,625,000
Area: 45,747 sq mi (118,484 sq km)
Pop. Density: 232/sq mi (90/sq km)
Capital: Lilongwe, 435,964

MALAYSIA
Official Name: Malaysia
Population: 22,445,000
Area: 127,320 sq mi (329,758 sq km)
Pop. Density: 176/sq mi (68/sq km)
Capital: Kuala Lumpur (de facto), 1,297,526
and Putrajaya (future)

MALDIVES
Official Name: Republic of Maldives
Population: 315,000
Area: 115 sq mi (298 sq km)
Pop. Density: 2,737/sq mi (1,057/sq km)
Capital: Male', 55,130

MALI
Official Name: Republic of Mali
Population: 11,170,000
Area: 482,077 sq mi (1,248,574 sq km)
Pop. Density: 23/sq mi (8.9/sq km)
Capital: Bamako, 658,275

MALTA
Official Name: Republic of Malta
Population: 395,000
Area: 122 sq mi (316 sq km)
Pop. Density: 3,238/sq mi (1,250/sq km)
Capital: Valletta, 7,262

MARSHALL ISLANDS
Official Name: Republic of the Marshall Islands
Population: 72,000
Area: 70 sq mi (181 sq km)
Pop. Density: 1,029/sq mi (397/sq km)
Capital: Majuro (island)

MAURITANIA
Official Name: Islamic Republic of Mauritania
Population: 2,790,000
Area: 397,956 sq mi (1,030,700 sq km)
Pop. Density: 7.0/sq mi (2.7/sq km)
Capital: Nouakchott, 393,325

MAURITIUS
Official Name: Republic of Mauritius
Population: 1,195,000
Area: 788 sq mi (2,040 sq km)
Pop. Density: 1,517/sq mi (586/sq km)
Capital: Port Louis, 132,460

Countries and Flags *continued*

MEXICO
Official Name: United Mexican States
Population: 102,640,000
Area: 758,452 sq mi (1,964,382 sq km)
Pop. Density: 135/sq mi (52/sq km)
Capital: Mexico City (Ciudad de México),
  8,489,007

MICRONESIA, FEDERATED STATES OF
Official Name: Federated States of
  Micronesia
Population: 135,000
Area: 271 sq mi (702 sq km)
Pop. Density: 498/sq mi (192/sq km)
Capital: Palikir, 5,047

MOLDOVA
Official Name: Republic of Moldova
Population: 4,435,000
Area: 13,012 sq mi (33,700 sq km)
Pop. Density: 341/sq mi (132/sq km)
Capital: Chişinău (Kishinev), 676,700

MONACO
Official Name: Principality of Monaco
Population: 32,000
Area: 0.8 sq mi (2.0 sq km)
Pop. Density: 40,000/sq mi (16,000/sq km)
Capital: Monaco, 32,000

MONGOLIA
Official Name: Mongolia
Population: 2,675,000
Area: 604,829 sq mi (1,566,500 sq km)
Pop. Density: 4.4/sq mi (1.7/sq km)
Capital: Ulan Bator (Ulaanbaatar), 616,900

MOROCCO
Official Name: Kingdom of Morocco
Population: 30,905,000
Area: 172,414 sq mi (446,550 sq km)
Pop. Density: 179/sq mi (69/sq km)
Capital: Rabat, 717,000

MOZAMBIQUE
Official Name: Republic of Mozambique
Population: 19,495,000
Area: 308,642 sq mi (799,380 sq km)
Pop. Density: 63/sq mi (24/sq km)
Capital: Maputo, 966,837

MYANMAR
Official Name: Union of Myanmar
Population: 42,120,000
Area: 261,228 sq mi (676,577 sq km)
Pop. Density: 161/sq mi (62/sq km)
Capital: Rangoon (Yangon), 2,705,053

NAMIBIA
Official Name: Republic of Namibia
Population: 1,810,000
Area: 317,818 sq mi (823,144 sq km)
Pop. Density: 5.7/sq mi (2.2/sq km)
Capital: Windhoek, 147,056

NAURU
Official Name: Republic of Nauru
Population: 12,000
Area: 8.0 sq mi (21 sq km)
Pop. Density: 1,481/sq mi (571/sq km)
Capital: Yaren District

NEPAL
Official Name: Kingdom of Nepal
Population: 25,580,000
Area: 56,827 sq mi (147,181 sq km)
Pop. Density: 450/sq mi (174/sq km)
Capital: Kathmandu, 421,258

NETHERLANDS
Official Name: Kingdom of the Netherlands
Population: 16,025,000
Area: 16,164 sq mi (41,864 sq km)
Pop. Density: 991/sq mi (383/sq km)
Capital: Amsterdam (designated), 727,053,
  and The Hague ('s-Gravenhage) (seat of
  government), 440,743

NEW ZEALAND
Official Name: New Zealand
Population: 3,885,000
Area: 104,454 sq mi (270,534 sq km)
Pop. Density: 37/sq mi (14/sq km)
Capital: Wellington, 150,301

NICARAGUA
Official Name: Republic of Nicaragua
Population: 4,970,000
Area: 50,054 sq mi (129,640 sq km)
Pop. Density: 99/sq mi (38/sq km)
Capital: Managua, 864,201

NIGER
Official Name: Republic of Niger
Population: 10,495,000
Area: 489,192 sq mi (1,267,000 sq km)
Pop. Density: 21/sq mi (8.3/sq km)
Capital: Niamey, 392,165

NIGERIA
Official Name: Federal Republic of Nigeria
Population: 128,285,000
Area: 356,669 sq mi (923,768 sq km)
Pop. Density: 360/sq mi (139/sq km)
Capital: Abuja, 250,000

NIUE
Official Name: Niue
Population: 2,000
Area: 100 sq mi (259 sq km)
Pop. Density: 20/sq mi (7.7/sq km)
Capital: Alofi, 682

NORTHERN MARIANA ISLANDS
Official Name: Commonwealth of the
  Northern Mariana Islands
Population: 76,000
Area: 184 sq mi (477 sq km)
Pop. Density: 413/sq mi (159/sq km)
Capital: Saipan (island)

NORWAY
Official Name: Kingdom of Norway
Population: 4,515,000
Area: 149,405 sq mi (386,958 sq km)
Pop. Density: 30/sq mi (12/sq km)
Capital: Oslo, 504,040

OMAN
Official Name: Sultanate of Oman
Population: 2,665,000
Area: 82,030 sq mi (212,457 sq km)
Pop. Density: 32/sq mi (13/sq km)
Capital: Muscat (Masqat), 34,683

PAKISTAN
Official Name: Islamic Republic of Pakistan
Population: 146,145,000
Area: 339,732 sq mi (879,902 sq km)
Pop. Density: 430/sq mi (166/sq km)
Capital: Islāmābād, 204,364

PALAU
Official Name: Republic of Palau
Population: 19,000
Area: 196 sq mi (508 sq km)
Pop. Density: 97/sq mi (37/sq km)
Capital: Koror (de facto), 9,018, and
Melekeok (future), 42

PANAMA
Official Name: Republic of Panama
Population: 2,865,000
Area: 29,157 sq mi (75,517 sq km)
Pop. Density: 98/sq mi (38/sq km)
Capital: Panamá, 415,964

PAPUA NEW GUINEA
Official Name: Independent State of Papua
New Guinea
Population: 5,110,000
Area: 178,704 sq mi (462,840 sq km)
Pop. Density: 29/sq mi (11/sq km)
Capital: Port Moresby, 246,664

PARAGUAY
Official Name: Republic of Paraguay
Population: 5,810,000
Area: 157,048 sq mi (406,752 sq km)
Pop. Density: 37/sq mi (14/sq km)
Capital: Asunción, 502,426

PERU
Official Name: Republic of Peru
Population: 27,720,000
Area: 496,225 sq mi (1,285,216 sq km)
Pop. Density: 56/sq mi (22/sq km)
Capital: Lima, 371,122

PHILIPPINES
Official Name: Republic of the Philippines
Population: 83,685,000
Area: 115,831 sq mi (300,000 sq km)
Pop. Density: 722/sq mi (279/sq km)
Capital: Manila, 1,654,761

POLAND
Official Name: Republic of Poland
Population: 38,630,000
Area: 120,728 sq mi (312,685 sq km)
Pop. Density: 320/sq mi (124/sq km)
Capital: Warsaw (Warszawa), 1,615,369

PORTUGAL
Official Name: Portuguese Republic
Population: 10,075,000
Area: 35,516 sq mi (91,985 sq km)
Pop. Density: 284/sq mi (110/sq km)
Capital: Lisbon (Lisboa), 663,394

PUERTO RICO
Official Name: Commonwealth of Puerto
Rico
Population: 3,950,000
Area: 3,515 sq mi (9,104 sq km)
Pop. Density: 1,124/sq mi (434/sq km)
Capital: San Juan, 421,958

QATAR
Official Name: State of Qatar
Population: 780,000
Area: 4,412 sq mi (11,427 sq km)
Pop. Density: 177/sq mi (68/sq km)
Capital: Ad Dawḩah (Doha), 361,540

ROMANIA
Official Name: Romania
Population: 22,340,000
Area: 91,699 sq mi (237,500 sq km)
Pop. Density: 244/sq mi (94/sq km)
Capital: Bucharest (Bucureşti), 2,067,545

RUSSIA
Official Name: Russian Federation
Population: 145,215,000
Area: 6,592,849 sq mi (17,075,400 sq km)
Pop. Density: 22/sq mi (8.5/sq km)
Capital: Moscow (Moskva), 8,368,449

RWANDA
Official Name: Republic of Rwanda
Population: 7,355,000
Area: 10,169 sq mi (26,338 sq km)
Pop. Density: 723/sq mi (279/sq km)
Capital: Kigali, 232,733

ST. KITTS AND NEVIS
Official Name: Federation of St. Kitts and Nevis
Population: 39,000
Area: 104 sq mi (269 sq km)
Pop. Density: 375/sq mi (145/sq km)
Capital: Basseterre, 11,295

ST. LUCIA
Official Name: St. Lucia
Population: 160,000
Area: 238 sq mi (616 sq km)
Pop. Density: 672/sq mi (260/sq km)
Capital: Castries, 11,147

Countries and Flags *continued*

ST. VINCENT AND THE GRENADINES
Official Name: St. Vincent and the
Grenadines
Population: 115,000
Area: 150 sq mi (388 sq km)
Pop. Density: 767/sq mi (296/sq km)
Capital: Kingstown, 15,466

SAMOA
Official Name: Independent State of Samoa
Population: 180,000
Area: 1,093 sq mi (2,831 sq km)
Pop. Density: 165/sq mi (64/sq km)
Capital: Apia, 34,126

SAN MARINO
Official Name: Republic of San Marino
Population: 28,000
Area: 24 sq mi (61 sq km)
Pop. Density: 1,186/sq mi (459/sq km)
Capital: San Marino, 2,794

SAO TOME AND PRINCIPE
Official Name: Democratic Republic of Sao
Tome and Principe
Population: 170,000
Area: 372 sq mi (964 sq km)
Pop. Density: 457/sq mi (176/sq km)
Capital: São Tomé, 5,245

SAUDI ARABIA
Official Name: Kingdom of Saudi Arabia
Population: 23,130,000
Area: 830,000 sq mi (2,149,690 sq km)
Pop. Density: 28/sq mi (11/sq km)
Capital: Riyadh (Ar Riyād), 1,250,000

SENEGAL
Official Name: Republic of Senegal
Population: 10,435,000
Area: 75,951 sq mi (196,712 sq km)
Pop. Density: 137/sq mi (53/sq km)
Capital: Dakar, 1,490,450

SERBIA AND MONTENEGRO
Official Name: Serbia and Montenegro
Population: 10,665,000
Area: 39,449 sq mi (102,173 sq km)
Pop. Density: 270/sq mi (104/sq km)
Capital: Belgrade (Beograd), 1,136,786

SEYCHELLES
Official Name: Republic of Seychelles
Population: 80,000
Area: 176 sq mi (455 sq km)
Pop. Density: 455/sq mi (176/sq km)
Capital: Victoria, 24,907

SIERRA LEONE
Official Name: Republic of Sierra Leone
Population: 5,525,000
Area: 27,925 sq mi (72,325 sq km)
Pop. Density: 198/sq mi (76/sq km)
Capital: Freetown, 469,776

SINGAPORE
Official Name: Republic of Singapore
Population: 4,375,000
Area: 246 sq mi (636 sq km)
Pop. Density: 17,814/sq mi (6,879/sq km)
Capital: Singapore, 4,017,700

SLOVAKIA
Official Name: Slovak Republic
Population: 5,420,000
Area: 18,933 sq mi (49,035 sq km)
Pop. Density: 286/sq mi (111/sq km)
Capital: Bratislava, 441,453

SLOVENIA
Official Name: Republic of Slovenia
Population: 1,930,000
Area: 7,821 sq mi (20,256 sq km)
Pop. Density: 247/sq mi (95/sq km)
Capital: Ljubljana, 292,589

SOLOMON ISLANDS
Official Name: Solomon Islands
Population: 490,000
Area: 10,954 sq mi (28,370 sq km)
Pop. Density: 45/sq mi (17/sq km)
Capital: Honiara, 30,413

SOMALIA
Official Name: Somalia
Population: 7,620,000
Area: 246,201 sq mi (637,657 sq km)
Pop. Density: 31/sq mi (12/sq km)
Capital: Mogadishu (Muqdisho), 600,000

SOUTH AFRICA
Official Name: Republic of South Africa
Population: 43,645,000
Area: 470,693 sq mi (1,219,090 sq km)
Pop. Density: 93/sq mi (36/sq km)
Capital: Pretoria (administrative), 525,583,
Cape Town (legislative), 854,616, and
Bloemfontein (judicial), 126,867

SPAIN
Official Name: Kingdom of Spain
Population: 40,060,000
Area: 194,885 sq mi (504,750 sq km)
Pop. Density: 206/sq mi (79/sq km)
Capital: Madrid, 2,882,860

SRI LANKA
Official Name: Democratic Socialist Republic
of Sri Lanka
Population: 19,495,000
Area: 24,962 sq mi (64,652 sq km)
Pop. Density: 781/sq mi (302/sq km)
Capital: Colombo (designated), 612,000,
and Sri Jayewardenepura Kotte (seat of
government), 108,000

SUDAN
Official Name: Republic of the Sudan
Population: 36,585,000
Area: 967,500 sq mi (2,505,813 sq km)
Pop. Density: 38/sq mi (15/sq km)
Capital: Khartoum (Al Kharṭum), 473,597

SURINAME
Official Name: Republic of Suriname
Population: 435,000
Area: 63,251 sq mi (163,820 sq km)
Pop. Density: 6.9/sq mi (2.7/sq km)
Capital: Paramaribo, 241,000

SWAZILAND
Official Name: Kingdom of Swaziland
Population: 1,115,000
Area: 6,704 sq mi (17,364 sq km)
Pop. Density: 166/sq mi (64/sq km)
Capital: Mbabane (administrative), 38,290,
and Lobamba (legislative)

SWEDEN
Official Name: Kingdom of Sweden
Population: 8,875,000
Area: 173,732 sq mi (449,964 sq km)
Pop. Density: 51/sq mi (20/sq km)
Capital: Stockholm, 674,452

SWITZERLAND
Official Name: Swiss Confederation
Population: 7,295,000
Area: 15,943 sq mi (41,293 sq km)
Pop. Density: 458/sq mi (177/sq km)
Capital: Bern (Berne), 136,338

SYRIA
Official Name: Syrian Arab Republic
Population: 16,940,000
Area: 71,498 sq mi (185,180 sq km)
Pop. Density: 237/sq mi (91/sq km)
Capital: Damascus (Dimashq), 1,549,932

TAIWAN
Official Name: Republic of China
Population: 22,460,000
Area: 13,901 sq mi (36,002 sq km)
Pop. Density: 1,616/sq mi (624/sq km)
Capital: T'aipei, 2,706,453

TAJIKISTAN
Official Name: Republic of Tajikistan
Population: 6,650,000
Area: 55,251 sq mi (143,100 sq km)
Pop. Density: 120/sq mi (46/sq km)
Capital: Dushanbe, 582,400

TANZANIA
Official Name: United Republic of Tanzania
Population: 36,705,000
Area: 364,900 sq mi (945,087 sq km)
Pop. Density: 101/sq mi (39/sq km)
Capital: Dar es Salaam (de facto), 1,096,000,
and Dodoma (legislative), 85,000

THAILAND
Official Name: Kingdom of Thailand
Population: 62,080,000
Area: 198,115 sq mi (513,115 sq km)
Pop. Density: 313/sq mi (121/sq km)
Capital: Bangkok (Krung Thep), 5,620,591

TOGO
Official Name: Republic of Togo
Population: 5,220,000
Area: 21,925 sq mi (56,785 sq km)
Pop. Density: 238/sq mi (92/sq km)
Capital: Lomé, 500,000

TONGA
Official Name: Kingdom of Tonga
Population: 105,000
Area: 288 sq mi (747 sq km)
Pop. Density: 364/sq mi (141/sq km)
Capital: Nuku'alofa, 22,400

TRINIDAD AND TOBAGO
Official Name: Republic of Trinidad and
Tobago
Population: 1,165,000
Area: 1,980 sq mi (5,128 sq km)
Pop. Density: 588/sq mi (227/sq km)
Capital: Port of Spain, 50,878

TUNISIA
Official Name: Republic of Tunisia
Population: 9,760,000
Area: 63,170 sq mi (163,610 sq km)
Pop. Density: 155/sq mi (60/sq km)
Capital: Tunis, 674,142

TURKEY
Official Name: Republic of Turkey
Population: 66,905,000
Area: 302,541 sq mi (783,577 sq km)
Pop. Density: 221/sq mi (85/sq km)
Capital: Ankara, 2,559,471

TURKMENISTAN
Official Name: Turkmenistan
Population: 4,645,000
Area: 188,457 sq mi (488,100 sq km)
Pop. Density: 25/sq mi (10/sq km)
Capital: Ashgabat, 412,200

TUVALU
Official Name: Tuvalu
Population: 11,000
Area: 10 sq mi (26 sq km)
Pop. Density: 1,100/sq mi (423/sq km)
Capital: Funafuti, 2,191

UGANDA
Official Name: Republic of Uganda
Population: 24,335,000
Area: 93,104 sq mi (241,139 sq km)
Pop. Density: 261/sq mi (101/sq km)
Capital: Kampala, 773,463

Countries and Flags *continued*

UKRAINE
Official Name: Ukraine
Population: 48,570,000
Area: 233,090 sq mi (603,700 sq km)
Pop. Density: 208/sq mi (80/sq km)
Capital: Kiev (Kyyiv), 2,630,000

UNITED ARAB EMIRATES
Official Name: United Arab Emirates
Population: 2,425,000
Area: 32,278 sq mi (83,600 sq km)
Pop. Density: 75/sq mi (29/sq km)
Capital: Abū Ẓaby (Abu Dhabi), 242,975

UNITED KINGDOM
Official Name: United Kingdom of Great
  Britain and Northern Ireland
Population: 59,715,000
Area: 94,249 sq mi (244,101 sq km)
Pop. Density: 634/sq mi (245/sq km)
Capital: London, 7,650,944

UNITED STATES
Official Name: United States of America
Population: 279,310,000
Area: 3,717,796 sq mi (9,629,091 sq km)
Pop. Density: 75/sq mi (29/sq km)
Capital: Washington, D.C., 572,059

URUGUAY
Official Name: Oriental Republic of Uruguay
Population: 3,375,000
Area: 68,500 sq mi (177,414 sq km)
Pop. Density: 49/sq mi (19/sq km)
Capital: Montevideo, 1,303,182

UZBEKISTAN
Official Name: Republic of Uzbekistan
Population: 25,355,000
Area: 172,742 sq mi (447,400 sq km)
Pop. Density: 147/sq mi (57/sq km)
Capital: Tashkent, 2,113,300

VANUATU
Official Name: Republic of Vanuatu
Population: 195,000
Area: 4,707 sq mi (12,190 sq km)
Pop. Density: 41/sq mi (16/sq km)
Capital: Port Vila, 19,311

VATICAN CITY
Official Name: State of the Vatican City
Population: 1,000
Area: 0.2 sq mi (0.4 sq km)
Pop. Density: 5,000/sq mi (2,500/sq km)
Capital: Vatican City, 1,000

VENEZUELA
Official Name: Republic of Venezuela
Population: 24,105,000
Area: 352,145 sq mi (912,050 sq km)
Pop. Density: 68/sq mi (26/sq km)
Capital: Caracas, 1,822,465

VIETNAM
Official Name: Socialist Republic of Vietnam
Population: 80,520,000
Area: 127,428 sq mi (330,036 sq km)
Pop. Density: 632/sq mi (244/sq km)
Capital: Hanoi, 905,939

YEMEN
Official Name: Republic of Yemen
Population: 18,385,000
Area: 203,850 sq mi (527,968 sq km)
Pop. Density: 90/sq mi (35/sq km)
Capital: Sanʻāʼ, 427,150

ZAMBIA
Official Name: Republic of the Zambia
Population: 9,865,000
Area: 290,586 sq mi (752,614 sq km)
Pop. Density: 34/sq mi (13/sq km)
Capital: Lusaka, 982,362

ZIMBABWE
Official Name: Republic of Zimbabwe
Population: 11,375,000
Area: 150,873 sq mi (390,759 sq km)
Pop. Density: 75/sq mi (29/sq km)
Capital: Harare (Salisbury), 1,189,103

Introduction to the Maps and Legend

Continental and regional coverage of the world's land areas is provided by the following section of thematic maps and physical-political reference maps. The reference map section falls into a continental arrangement: North America, South America, Europe, Africa, Asia, and Oceania. Introducing each regional reference map section are several basic thematic maps.

To aid the reader in understanding the relative sizes of continents and of some of the countries and regions, uniform scales for comparable areas were used as far as possible. Most of the world is covered by a series of regional maps at scales of 1:16,000,000 and 1:12,000,000. Maps at 1:10,000,000 provide even greater detail for parts of Europe. The United States and parts of South America are mapped at 1:4,000,000.

Many of the symbols used are self-explanatory. A complete legend below provides a key to the symbols on the reference maps in the atlas.

The color tints on the maps depict the varying elevations and depths of land areas and bodies of water. The Relief legend that accompanies each map shows the specific elevation or depth that each color tint represents.

The surface configuration is represented by hill-shading, which gives the three-dimensional impression of landforms. This terrain representation is superimposed on the layer tints to convey a realistic and readily visualized impression of the surface. The combination of altitudinal tints and hill-shading best shows elevation, relief, steepness of slope, and ruggedness of terrain.

If the world used one alphabet and language, no particular difficulty would arise in understanding place-names. However, some of the people of the world, the Chinese and the Japanese, for example, use non-alphabetic languages. Their symbols are transliterated into the Roman alphabet. In this atlas, a "local-name" policy generally was used for naming cities, towns, and all local topographic and water features. However, for a few major cities the Anglicized name was preferred and the local name given in parentheses: for instance, Moscow (Moskva), Vienna (Wien), Bangkok (Krung Thep). In countries where more than one official language is used, a name appears in the dominant local language. The generic parts of local names for topographic and water features are self-explanatory in many cases because of the associated map symbols or type styles. A complete list of foreign generic names is given in the Glossary.

Physical-Political Reference Map Legend

Cultural Features

Political Boundaries

International (Demarcated, Undemarcated, and Administrative) (over water)
Disputed de facto
Claim Boundary
Indefinite or Undefined
Secondary, State, Provincial, etc. (over water)
Parks, Indian Reservations
City Limits — Urbanized Areas
Neighborhoods, Sections of City

Populated Places
◉ 1,000,000 and over
◎ 250,000 to 1,000,000
⊙ 100,000 to 250,000
• 25,000 to 100,000
○ 0 to 25,000
TŌKYŌ National Capitals
Boise Secondary Capitals

Note: On maps at 1:20,000,000 and smaller the town symbols do not follow the specific population classification shown above. On all maps, type size indicates the relative importance of the city.

Transportation

Railroads
Railroads On 1:1,000,000 scale maps
Railroad Ferries
Roads
Major / Other On 1:1,000,000 scale maps
Major / Other On 1:4,000,000 scale maps
On other scale maps
Caravan Routes
✈ Airports

Other Cultural Features
Dams
Pipelines
▲ Points of Interest
Ruins

Land Features
△ Peaks, Spot Heights
Passes
Sand
Contours

Water Features

Lakes and Reservoirs
Fresh Water
Fresh Water: Intermittent
Salt Water
Salt Water: Intermittent

Other Water Features
Salt Basins, Flats
Swamps
Ice Caps and Glaciers
Rivers
Intermittent Rivers
Aqueducts and Canals
Ship Channels
Falls
Rapids
Springs
△ Water Depths
Fishing Banks
Sand Bars
Reefs

Note: Country populations used throughout the atlas are 2002 estimates based on 2001 U.S. Census Bureau figures and other available information. City populations in the continent "At a Glance" sections reflect the latest available official data.

ARCTIC OCEAN

NOVOSIBIRSKIJE OSTROVA

ZEML'A FRANCA-IOSIFA

More Laptevych

75°

/ALBARD (Nor.)

Barents Sea

NOVAJA ZEML'A

Karskoje More

Arctic Circle

60°

B

Noril'sk

RWAY

SWEDEN

FINLAND

Archangel'sk

Jenisej

Lena

Jakutsk

Anadyr

Helsinki

Oslo

SANKT-PETERBURG

R U S S I A

Stockholm

EST.

LAT.

LITH.

Nižnij Novgorod

MOSKVA

Jekaterinburg

Novosibirsk

Sea of Okhotsk

Bering Sea

ALEUTIAN IS. (U.S.)

C

60°

Ob'

Ozero Bajkal

OSTROV SACHALIN

Petropavlovsk-Kamcatskij

GERMANY

POLAND

BELA.

Kyjiv

K A Z A K H S T A N

A

L

T

A

J

MONGOLIA

Harbin

Sea of Japan

45°

EUROPE

AUS.

SLVK.

HUNG.

CRO.

SER.

UKRAINE

MOLD.

Volga

Caspian

Aral Sea

G

O

R

Y

G O B I

KOREA

SOUL

JAPAN

ŌSAKA

TŌKYŌ

D

Roma

ITALY

SWITZ.

ALB.

BUL.

ROM.

Gora El'brus
5633

Black Sea

GEORG.

UZBEK.

Taškent

KYRG.

C H I N A

Yellow Sea

SHANGHAI

30°

GREECE

TURKEY

Istanbul

ARM.

AZER.

TURKMENISTAN

TADŽ.

Xi'an

Wuhan

Mediterranean Sea

MALTA

SYRIA

LEB.

IRAQ

JORDAN

Tehrān

IRAN

AFGHANISTAN

H
I
M
A
L
A
Y
A
S

Chongqing

PACIFIC

TUNISIA

ISRAEL

KUWAIT

PAKISTAN

NEPAL

BHU.

Guangzhou

Xianggang
(Hong Kong)

TAIWAN

OCEAN

15°

E

RIA

LIBYA

AL-QĀHIRAH

CAIRO

EGYPT

SAUDI

QATAR

UNITED ARAB EMIRATES

OMAN

Karāchi

DELHI

Mount Everest
8848

BNGL.

KOLKATA

Tropic of Cancer

MYANMAR BURMA

LAOS

South

China

Philippine Sea

WAKE ISLAND (U.S.)

NIGER

CHAD

Al-Khartūm

SUDAN

Red Sea

ARABIA

YEMEN

Aden

MUMBAI

I N D I A

Chennai

Bay of

Bengal

THAILAND

VIETNAM

CAMB.

Krung Thep
Bangkok

Sea

MANILA

GUAM (U.S.)

F

AFRICA

NIGERIA

Lagos

CEN. AFR. REP.

DJIBOUTI

ETHIOPIA

SOMALIA

Muqdisho

Arabian

Sea

SRI LANKA

Colombo

PHILIPPINES

M
I
C
R
O
N
E
S
I
A

ATORIAL GUINEA

CAMEROON

UGANDA

KENYA

Nairobi

MALDIVES

BRUNEI

15°

GABON

CONGO

DEM. REP. OF THE CONGO

RWANDA

BURUNDI

Lake Victoria

Kilimanjaro 5895

Equator

MALAYSIA

Singapore

BORNEO

SULAWESI

Equator

KIRIBATI

0°

Kinshasa

ZAIRE

TANZANIA

SEYCHELLES

CHAGOS ARCHIPELAGO
(B.I.O.T.)

SUMATERA

IRIAN JAYA

PAPUA NEW GUINEA

SOLOMON ISLANDS

TUVALU

Luanda

JAKARTA

I N D O N E S I A

NEW GUINEA

SOLOMON ISLANDS

G

ANGOLA

I N D I A N

JAWA

EAST TIMOR

Port Moresby

VANUATU

ZAMBIA

MADAGASCAR

O C E A N

CHRISTMAS ISLAND
(Austl.)

TIMOR

M
E
L
A
N
E
S
I
A

15°

ZIMBABWE

MOZAMBIQUE

Mozambique Channel

MAURITIUS

Tropic of Capricorn

Cairns

Coral Sea

NEW CALEDONIA
(Fr.)

FIJI

H

NAMIBIA

BOTSWANA

REUNION
(Fr.)

A U S T R A L I A

Brisbane

Johannesburg

SWAZILAND

Perth

30°

LESOTHO

Sydney

Cape Town

SOUTH AFRICA

Durban

Melbourne

Mount Kosciuszko
2230

Tasman Sea

NEW ZEALAND

I

CAPE OF GOOD HOPE

TASMANIA

Wellington

45°

ÎLES KERGUELEN
(F.S.A.T.)

J

60°

Antarctic Circle

K

ENDERBY LAND

WILKES LAND

75°

A C T I C A

L

90°

14 30° 15 45° 16 60° 17 75° 18 90° 19 105° 20 120° 21 135° 22 150° 23 165° 24 180°

Kilometers

0 1000 2000 3000 Km.

Miles

0 1000 3000 Mi.

Robinson Projection

World Physical

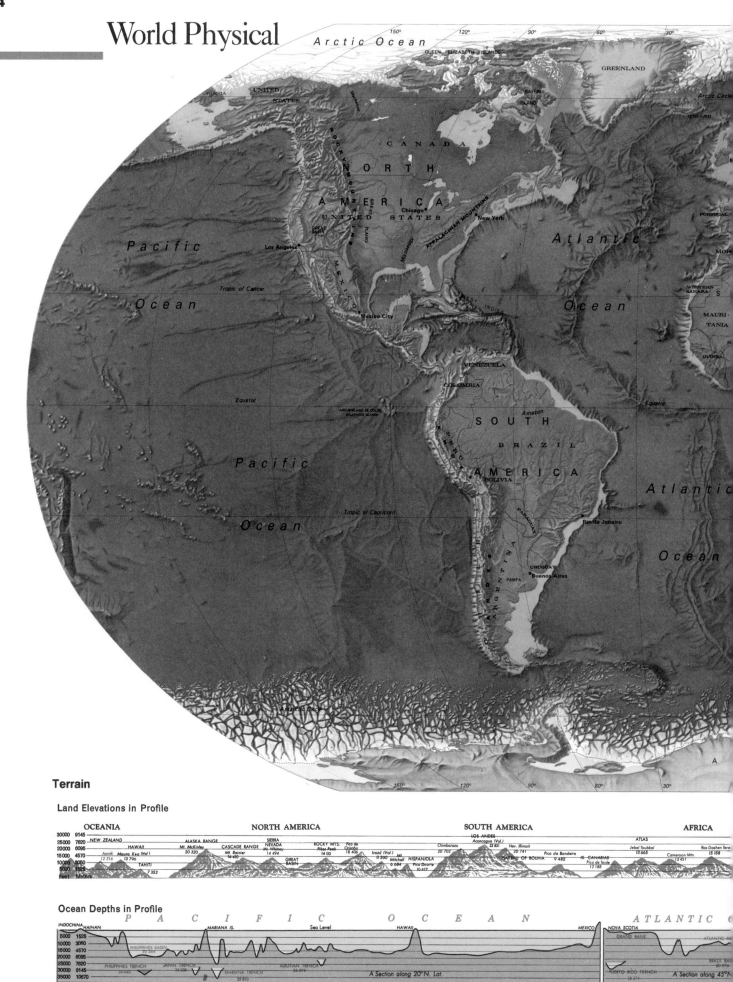

Arctic Ocean

QUEEN ELIZABETH ISLANDS

GREENLAND

RUSSIA

UNITED STATES

Arctic Circle

BAFFIN ISLAND

ICELAND

C A N A D A

N O R T H

A M E R I C A

Chicago

New York

PORTUGAL

Pacific

U N I T E D   S T A T E S

APPALACHIAN MOUNTAINS

Atlantic

GREAT BASIN

Los Angeles

Ocean

MOR.

Tropic of Cancer

WESTERN SAHARA

Ocean

S

MEXICO

MAURI-TANIA

Mexico City

WEST INDIES

GUINEA

Equator

Equator

VENEZUELA

COLOMBIA

ARCHIPIELAGO DE COLON
GALAPAGOS ISLANDS

Amazon

S O U T H

Pacific

PERU

B R A Z I L

A M E R I C A

Atlantic

BOLIVIA

Ocean

Tropic of Capricorn

PARAGUAY

Rio de Janeiro

Ocean

URUGUAY

PAMPA   Buenos Aires

A

Atlantic Ridge

150°   120°   90°   60°   30°

Terrain

Land Elevations in Profile

	OCEANIA	NORTH AMERICA	SOUTH AMERICA	AFRICA

30000 9145
25000 7620   NEW ZEALAND   ALASKA RANGE   SIERRA   LOS ANDES   ATLAS
20000 6095   HAWAII   Mt. McKinley CASCADE RANGE NEVADA   ROCKY MTS. Pico de   Chimborazo Aconcagua (Vol.) Nev. Illimani   Ras Dashen Tera.
                                   20 320   Mt. Rainier Mt. Whitney Pikes Peak Orizaba   20 702   22 831   20 741   Pico da Bandeira   Jebel Toubkal   15 158
15000 4570   Aoraki Mauna Kea (Vol.)   14 410   14 494   14 110   18 406   Irazú (Vol.) Mt.   IS. CANARIAS   Cameroon Mtn. 13 665
              12 316   13 796   11 200 Mitchell HISPANIOLA PLATEAU OF BOLIVIA 9 482   Pico de Teide   13 451
10000 3050   TAHITI   GREAT   6 684 Pico Duarte   12 188
5000   1525   BASIN   10 417
Feet   Meters   7 352

Ocean Depths in Profile

P A C I F I C   O C E A N   A T L A N T I C   O

INDOCHINA HAINAN   MARIANA IS.   Sea Level   HAWAII   MEXICO   NOVA SCOTIA
5000   1525   GRAND BANK   ATLANTIC
10000 3050   PHILIPPINES BASIN   BRAZIL BA.
15000 4570   20 564   20 79
20000 6095   A Section along 45°
25000 7620   PHILIPPINES TRENCH   JAPAN TRENCH   ALEUTIAN TRENCH   PUERTO RICO TRENCH
30000 9145   34 440   34 035   20 674   A Section along 20° N. Lat.   28 374
35000 10670   MARIANA TRENCH
                                        35 810
Feet   Meters

Elevations and depress

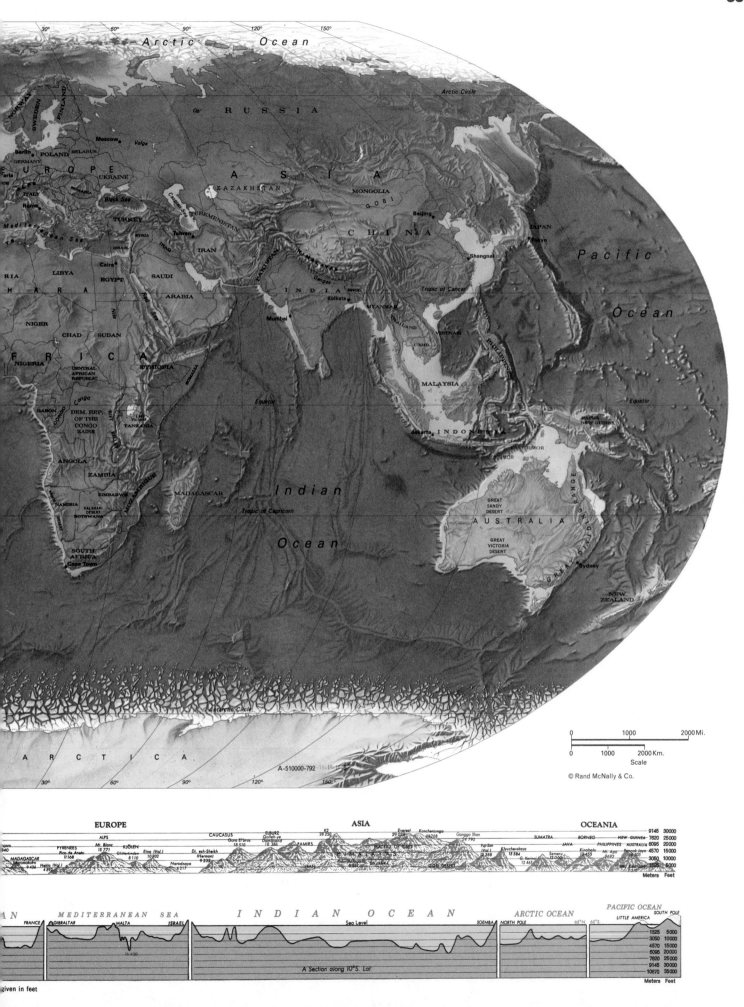

© Rand McNally & Co.

World Climate

© Rand McNally & Co.
N-ANS10000-C1- -1--1-2

World Climate Zones

Tropical

Tropical rain forest

Tropical savanna

Dry

Steppe

Desert

Temperate - Mild and rainy winter

Mediterranean

Humid subtropical

Marine West Coast

Temperate - Cold and snowy winter

Humid continental, warm summer

Humid continental, cool summer

Subarctic

Polar

Tundra

Ice cap

Highlands

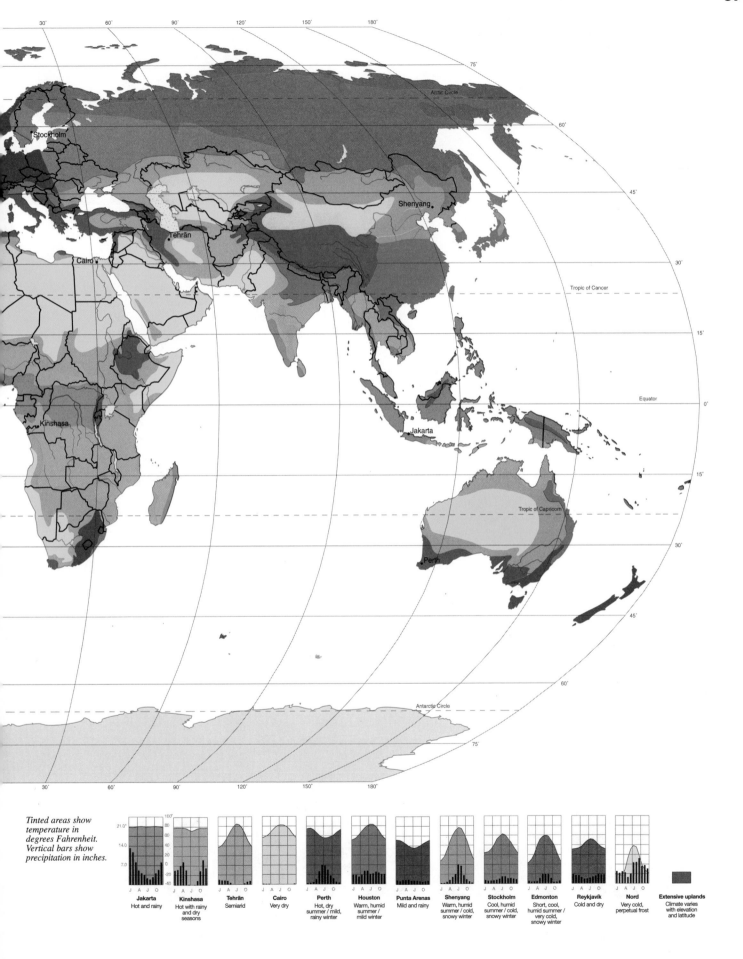

Tinted areas show temperature in degrees Fahrenheit. Vertical bars show precipitation in inches.

Jakarta
Hot and rainy

Kinshasa
Hot with rainy and dry seasons

Tehrān
Semiarid

Cairo
Very dry

Perth
Hot, dry summer / mild, rainy winter

Houston
Warm, humid summer / mild winter

Punta Arenas
Mild and rainy

Shenyang
Warm, humid summer / cold, snowy winter

Stockholm
Cool, humid summer / cold, snowy winter

Edmonton
Short, cool, humid summer / very cold, snowy winter

Reykjavík
Cold and dry

Nord
Very cold, perpetual frost

Extensive uplands
Climate varies with elevation and latitude

World Vegetation

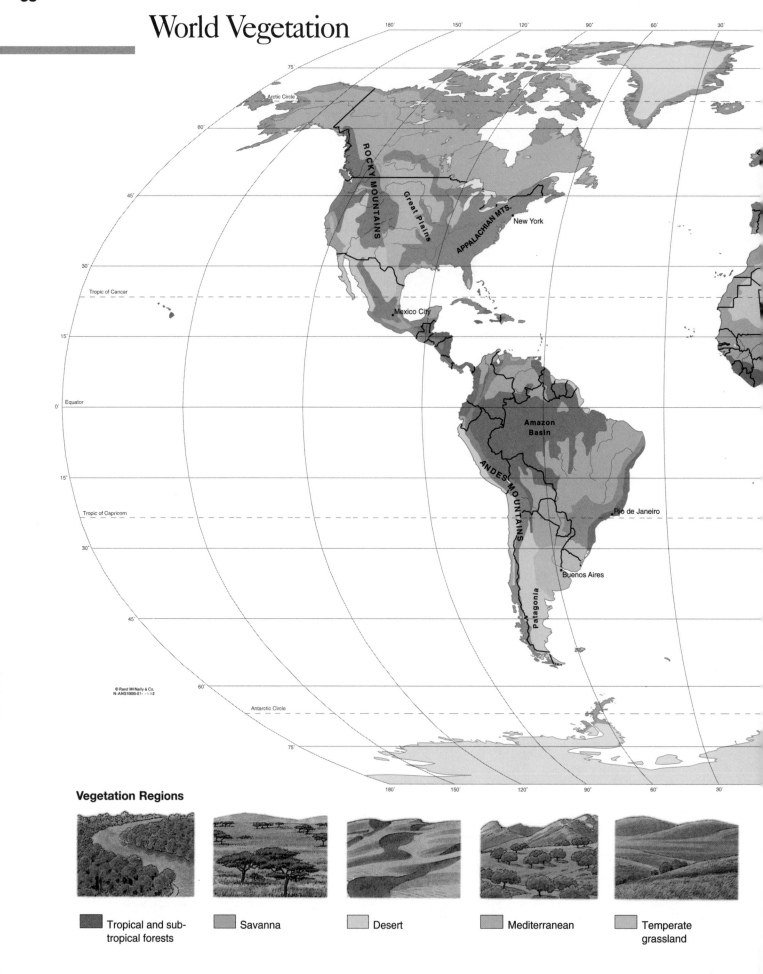

180° 150° 120° 90° 60° 30°

75°

Arctic Circle

60°

45°

ROCKY MOUNTAINS

Great Plains

APPALACHIAN MTS.

New York

30°

Tropic of Cancer

15°

Mexico City

Equator 0°

15°

Amazon Basin

ANDES MOUNTAINS

Tropic of Capricorn

30°

Rio de Janeiro

Buenos Aires

45°

Patagonia

© Rand McNally & Co.
N-ANS10000-E1- - ½-½2

60°

Antarctic Circle

75°

180° 150° 120° 90° 60° 30°

Vegetation Regions

Tropical and sub-tropical forests

Savanna

Desert

Mediterranean

Temperate grassland

| | Temperate forest | | Taiga (northern forests) | | Tundra (lichen and moss) | | Mountain | | Polar and high mountain |

World Population

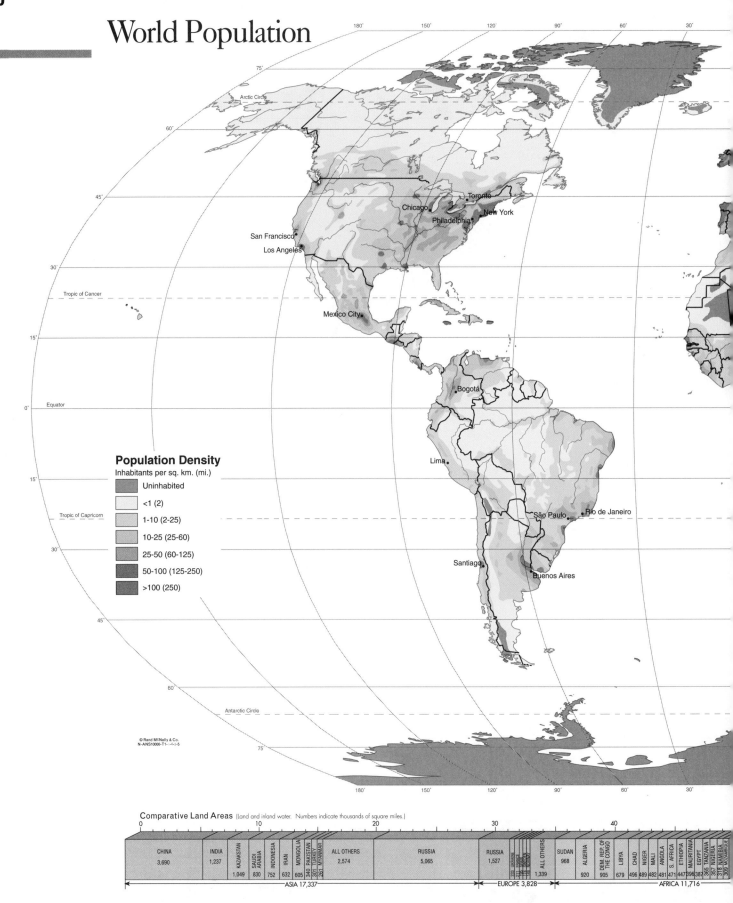

Population Density
Inhabitants per sq. km. (mi.)

	Uninhabited
	<1 (2)
	1-10 (2-25)
	10-25 (25-60)
	25-50 (60-125)
	50-100 (125-250)
	>100 (250)

© Rand McNally & Co.
N-ANS10000-T1-·-·-5

Comparative Land Areas (Land and inland water. Numbers indicate thousands of square miles.)

CHINA 3,690	INDIA 1,237	KAZAKHSTAN 1,049	SAUDI ARABIA 830	INDONESIA 752	IRAN 632	MONGOLIA 605	PAKISTAN 340	TURKEY 301	MYANMAR 261	ALL OTHERS 2,574	RUSSIA 5,065	RUSSIA 1,527	UKRAINE 233	FRANCE 211	SPAIN 195	SWEDEN 174	GERMANY 138	ALL OTHERS 1,339	SUDAN 968	ALGERIA 920	DEM. REP. OF THE CONGO 905	LIBYA 679	CHAD 496	NIGER 489	MALI 482	ANGOLA 481	S. AFRICA 471	MAURITANIA 447	ETHIOPIA 396	EGYPT 387	TANZANIA 365	NIGERIA 357	MOZAMBIQUE 309

ASIA 17,337 · EUROPE 3,828 · AFRICA 11,716

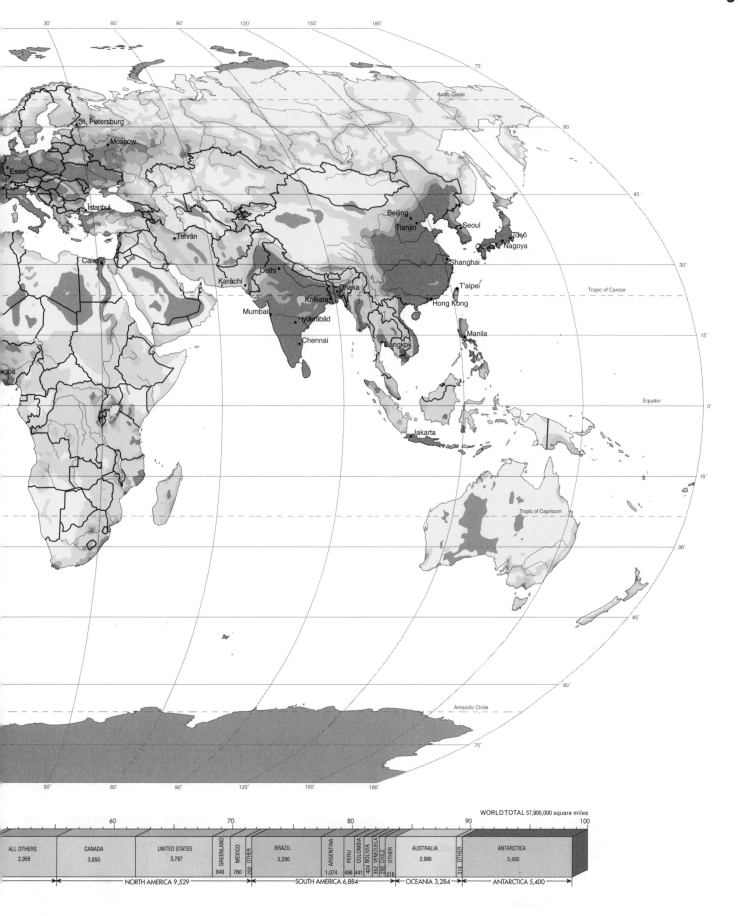

WORLD TOTAL 57,900,000 square miles

	60		70		80		90		100

| ALL OTHERS 2,956 | CANADA 3,850 | UNITED STATES 3,787 | GREENLAND 840 | MEXICO 760 | OTHER 292 | BRAZIL 3,286 | ARGENTINA 1,074 | PERU 496 | COLOMBIA 441 | BOLIVIA 424 | VENEZUELA 352 | CHILE 292 | OTHER 518 | AUSTRALIA 2,966 | OTHER 318 | ANTARCTICA 5,400 |

NORTH AMERICA 9,529 ── SOUTH AMERICA 6,884 ── OCEANIA 3,284 ── ANTARCTICA 5,400

World Environments

Environments

Urban	Cropland
Cropland, woodland	Cropland, grazing land
Grassland, grazing land	

© Rand McNally & Co.
N-ANS10000-M1- -:- ː-2

| | Forest, woodland | | Swamp, marsh | | Tundra | | Shrub, sparse grass, wasteland | | Barren land |

Time Zones

Standard time zone of even-numbered hours from Greenwich time	h m — hours, minutes
Standard time zone of odd-numbered hours from Greenwich time	
Time varies from the standard time zone by half an hour	
Time varies from the standard time zone by other than half an hour	

The standard time zone system, fixed by international agreement and by law in each country, is based on a theoretical division of the globe into 24 zones of 15° longitude each. The mid-meridian of each zone fixes the hour for the entire zone. The zero time zone extends 7½° east and 7½° west of the Greenwich meridian, 0° longitude. Since the earth rotates toward the east, time zones to the west of Greenwich are earlier, to the east, later.
Plus and minus hours at the top of the map are added to or subtracted from local time to find Greenwich time. Local standard time can be determined for any area in the world by adding one hour for each time zone counted in an easterly direction from one's own, or by subtracting one hour for each zone counted in a westerly direction. To separate one day from the next, the 180th meridian has been designated as the international date line. On both sides of the line the time of day is the same, but west of the line it is one day later than it is to the east. Countries that adhere to the international zone system adopt the zone applicable to their location. Some countries, however, establish time zones based on political boundaries, or adopt the time zone of a neighboring unit. For all or part of the year some countries also advance their time by one hour, thereby utilizing more daylight hours each day.

North America

North America is the world's third-largest continent, covering an area of 9.5 million square miles (24.7 million sq km). It lies primarily between the Arctic Circle and the Tropic of Cancer, and comes within 500 miles (800 km) of both the North Pole and the Equator. The continent's western flank is dominated by the spectacular Rocky Mountains. Covering vast stretches of the central United States and Canada are the fertile Great Plains, a large part of which is drained by the Mississippi River and its tributaries.

In the north, Hudson Bay is frozen for much of the year. Mexico, located in the continent's southern third, is mostly mountainous and dry, but farther south, the climate is wet. Many of the small Central American countries have volcanoes along the Pacific Coast.

North America at a glance

Land area: 9,500,000 square miles (24,700,000 sq km)

Estimated population: 488,780,000

Population density: 51/square mile (20/sq km)

Mean elevation: 2,000 feet (610 m)

Highest point: Mt. McKinley, Alaska, U.S. 20,230 feet (6,194 m)

Lowest point: Death Valley, California, U.S., 282 feet (86 m) below sea level

Longest river: Mississippi-Missouri, 3,740 mi (6,019 km)

Number of countries (incl. dependencies): 38

Largest independent country: Canada, 3,855,103 square miles (9,984,670 sq km)

Smallest independent country: St. Kitts and Nevis, 104 square miles (269 sq km)

Most populous independent country: United States, 279,310,000

Least populous independent country: St. Kitts and Nevis, 39,000

Largest city: Mexico City, pop. 8,489,007

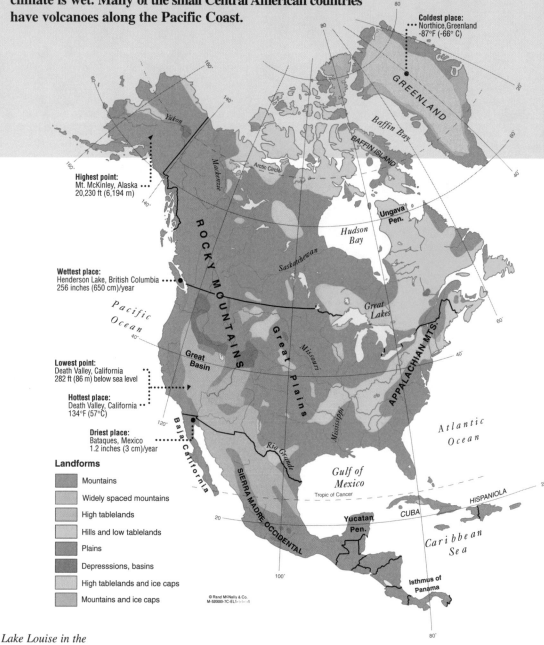

Coldest place: Northice, Greenland -87°F (-66° C)

Highest point: Mt. McKinley, Alaska 20,230 ft (6,194 m)

Wettest place: Henderson Lake, British Columbia 256 inches (650 cm)/year

Lowest point: Death Valley, California 282 ft (86 m) below sea level

Hottest place: Death Valley, California 134°F (57°C)

Driest place: Bataques, Mexico 1.2 inches (3 cm)/year

Landforms

- Mountains
- Widely spaced mountains
- High tablelands
- Hills and low tablelands
- Plains
- Depresssions, basins
- High tablelands and ice caps
- Mountains and ice caps

© Rand McNally & Co.
M-500000-7C-EL1-¦-¦- -¦

Lake Louise in the Canadian Rockies

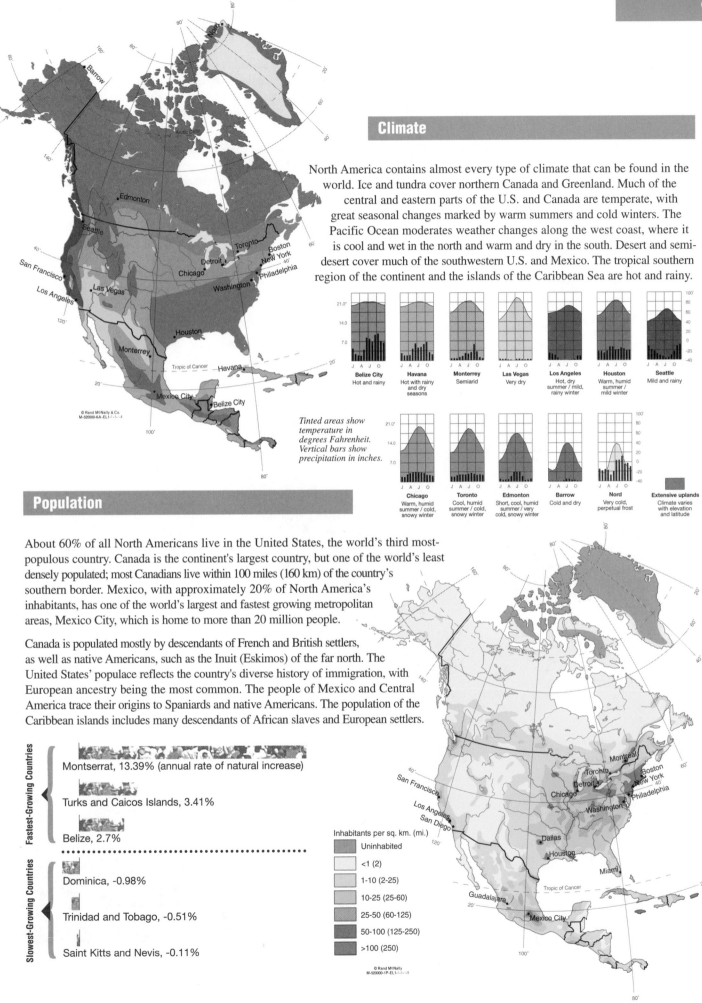

Climate

North America contains almost every type of climate that can be found in the world. Ice and tundra cover northern Canada and Greenland. Much of the central and eastern parts of the U.S. and Canada are temperate, with great seasonal changes marked by warm summers and cold winters. The Pacific Ocean moderates weather changes along the west coast, where it is cool and wet in the north and warm and dry in the south. Desert and semi-desert cover much of the southwestern U.S. and Mexico. The tropical southern region of the continent and the islands of the Caribbean Sea are hot and rainy.

Belize City Hot and rainy

Havana Hot with rainy and dry seasons

Monterrey Semiarid

Las Vegas Very dry

Los Angeles Hot, dry summer / mild, rainy winter

Houston Warm, humid summer / mild winter

Seattle Mild and rainy

Tinted areas show temperature in degrees Fahrenheit. Vertical bars show precipitation in inches.

Chicago Warm, humid summer / cold, snowy winter

Toronto Cool, humid summer / cold, snowy winter

Edmonton Short, cool, humid summer / very cold, snowy winter

Barrow Cold and dry

Nord Very cold, perpetual frost

Extensive uplands Climate varies with elevation and latitude

Population

About 60% of all North Americans live in the United States, the world's third most-populous country. Canada is the continent's largest country, but one of the world's least densely populated; most Canadians live within 100 miles (160 km) of the country's southern border. Mexico, with approximately 20% of North America's inhabitants, has one of the world's largest and fastest growing metropolitan areas, Mexico City, which is home to more than 20 million people.

Canada is populated mostly by descendants of French and British settlers, as well as native Americans, such as the Inuit (Eskimos) of the far north. The United States' populace reflects the country's diverse history of immigration, with European ancestry being the most common. The people of Mexico and Central America trace their origins to Spaniards and native Americans. The population of the Caribbean islands includes many descendants of African slaves and European settlers.

Fastest-Growing Countries

Montserrat, 13.39% (annual rate of natural increase)

Turks and Caicos Islands, 3.41%

Belize, 2.7%

Slowest-Growing Countries

Dominica, -0.98%

Trinidad and Tobago, -0.51%

Saint Kitts and Nevis, -0.11%

Inhabitants per sq. km. (mi.)

- Uninhabited
- <1 (2)
- 1-10 (2-25)
- 10-25 (25-60)
- 25-50 (60-125)
- 50-100 (125-250)
- >100 (250)

© Rand McNally
M-520000-1P-EL1-:-:-:-1

© Rand McNally & Co.
M-520000-6A-EL1-:-:-:-1

Environments and Land Use

Although only 12% of the continent is suitable for agriculture, North America is the world's leading food producer. Unlike other parts of the world, famine is virtually unknown. Large quantities of food, such as grains from the central U.S. and Canada, are exported worldwide. Sixteen percent of the continent is used for grazing, and the livestock raised on these lands are also an important source of food at home and abroad.

Forests cover one-third of the land, and the timber and paper industries are important to the U.S. and Canada. The continent has an extremely long coastline, and many countries send great fishing fleets to sea. This is espe cially true of Canada, whose eastern-most provinces are fittingly known as the "Maritime Provinces." However, sharp declines in catches due to over-fishing have put the industry in economic turmoil.

Blue waters, sunny skies, and idyllic beaches draw millions of tourists to the Caribbean each year. While tourism provides income for island countries that have few other assets, the economic gap between the visitors and the people who serve them remains dramatic.

One of the greatest challenges facing North America in the coming decades is a familiar one: coping with a growing population and dwindling resources. Pollution and the environment are divisive issues. Although the United States has begun to clean up its air and water, economic pressures will continue to be an argument for a relaxation of policies. Meanwhile, in Mexico, environmental issues have been pushed aside by concerns about the economy and the exploding population.

Urban

Cropland

Cropland and woodland

Cropland and grazing land

Grassland, grazing land

Forest, woodland

Swamp, marsh

Tundra

Shrub, sparse grass, wasteland

Barren land

© Rand McNally & Co.
N-ANS20000-M1-‑‑-‑‑-1

| 0 | 200 | 400 | 600 | 800 | 1000 Miles |

| 0 | 300 | 600 | 900 | 1200 | 1500 Kilometers |

*Field of corn in the
midwestern United States*

Urbanization in North America

Seen from the air, large portions of North America still bear the checkerboard imprint left by the people who settled the continent: a vast array of small farms that could be worked by one man and a horse. As industrialization swept through the United States in the second half of the 19th century, many farmers left their farms. Great numbers moved to fast-growing cities such as Chicago and St. Louis. Many small towns saw their populations dwindle.

In the cities, change was continuous. The former farm families were joined by waves of European immigrants. Over the next 100 years, urban populations con- tinued to grow, and cities that had once been separate became part of vast urban megalopolises. An example can be seen in the eastern United States, where a band of urban centers stretches almost continuously from Boston south through New York City to Washington D.C. (see map at right).

After World War II, as the U. S. middle class expanded, large numbers of fami- lies moved out of the city centers into suburban communities of mass-pro- duced, affordable homes. Many of those who remained were economically disad- vantaged. A shrinking tax base meant that cities could not support their infra- structures, and conditions in the inner cities worsened. In the suburbs the opposite was true: vast sums were spent building new roads, houses, and shopping centers.

The same process that transformed the U.S. is now occurring in Mexico. In 1945, 25% of the population was considered urban, but today the figure surpasses 70% (see graph above). Not content to lead lives as subsistence farmers, scores of Mexicans arrive in the cities each day in search of jobs that will enable them to provide a better life for their families.

Sadly, these dreams are often elusive. Most of the people end up in low-paying jobs, with their meager wages going to pay for the food that they once grew themselves. City officials are hard-pressed to provide clean water to their ever-growing populations, let alone electricity, trans- portation and education.

As elsewhere in the world, coping with the pressures of growing popu- lations is one of the greatest challenges facing North Americans today.

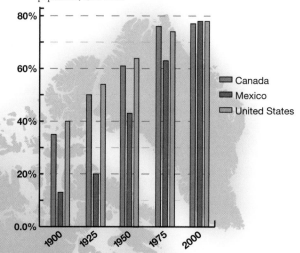

Rising Urban Population
Urban population as a percentage of total population, 1900-2000

Canada
Mexico
United States

(The definition of "urban" varies from country to country. In Canada, all towns with more than 1,000 people are considered urban, while in the U.S. and Mexico only towns with more than 2,500 people are defined as urban.)

Urban Centers

- 50,000-99,999 people
- 100,000-500,000 people
- Over 500,000 people

© Rand McNally
M-520000-9E-EL1-'-'- -'1

Mexico City, whose metropolitan area population exceeds 20 million, sprawls toward the distant mountains.

Scale 1:40 000 000; one inch to 630 miles. Lambert's Azimuthal Equal Area Projection
Elevations and depressions are given in feet

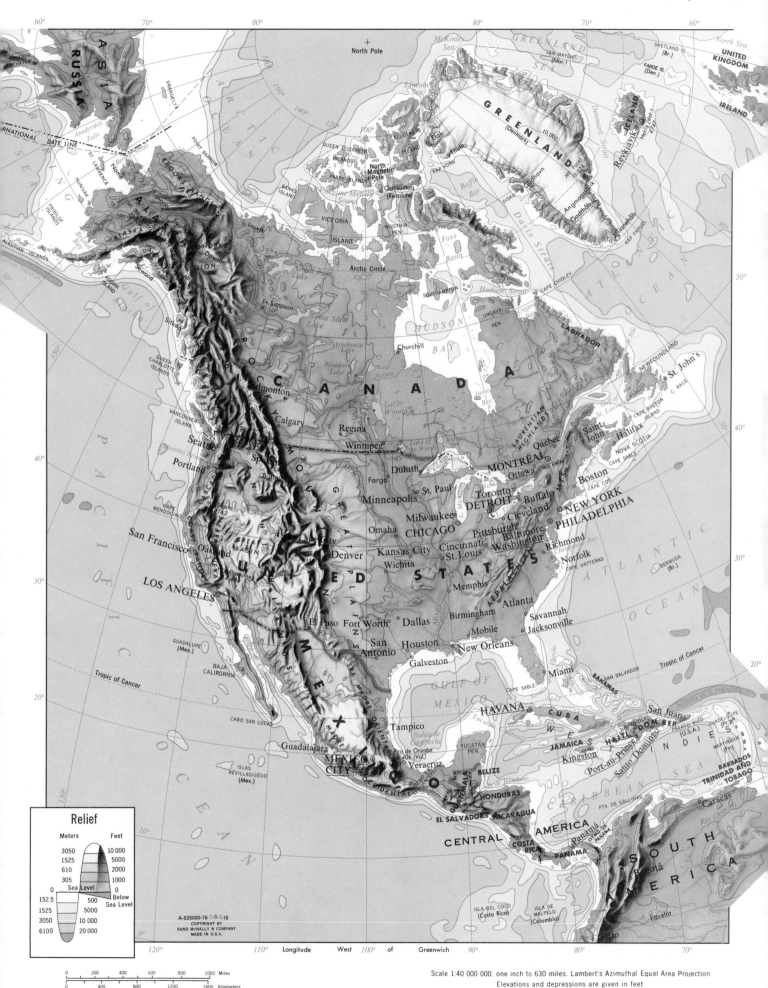

Relief

Meters	Feet
3050	10 000
1525	5000
610	2000
305	1000
0 Sea Level	Sea Level
152.5	500
1525	5000
3050	10 000
6100	20 000

Below Sea Level

A-520000-76 8-5-L-18
COPYRIGHT BY
RAND McNALLY & COMPANY
MADE IN U.S.A.

0 200 400 600 800 1000 Miles
0 400 800 1200 1600 Kilometers

Scale 1:40 000 000; one inch to 630 miles. Lambert's Azimuthal Equal Area Projection
Elevations and depressions are given in feet

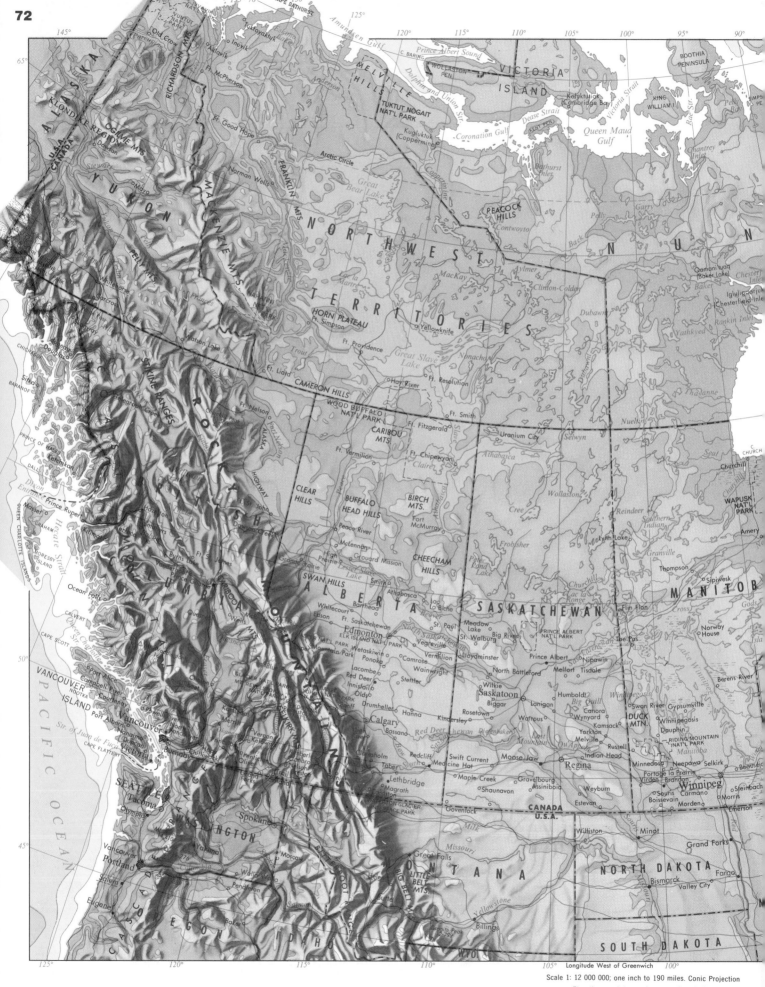

Scale 1: 12 000 000; one inch to 190 miles. Conic Projection
Elevations and depressions are given in feet

Longitude West of Greenwich

a

Same scale as main map

NEWFOUNDLAND AND LABRADOR

QUEBEC

CAPE BAULD

Gulf of St. Lawrence

GROS MORNE NAT'L PARK

Deer Lake

Corner Brook
Stephenville

C. ST. GEORGE

ST. JOHN

Twillingate

Botwood

Windsor
Grand Falls

Gander

Bonavista

TERRA NOVA NAT'L PARK

NEWFOUNDLAND

Trinity

CAPE RAY

Channel-Port-aux-Basques

CABOT STRAIT

CAPE NORTH

CAPE BRETON ISLAND

Grand Bank

Fortune Bay

Burin

St. John's

Placentia Bay

ST. PIERRE AND MIQUELON (Fr.)

ATLANTIC OCEAN

60° Longitude West of Greenwich 55°

MELVILLE PENINSULA

FOXE BASIN

PRINCE CHARLES ISLAND

BAFFIN ISLAND NAT'L PARK

BAFFIN ISLAND

Cumberland Sound

C. MERCY

SOUTHAMPTON ISLAND

Arctic Circle

FOXE PEN.

FOXE CHANNEL

Iqaluit

EVERETT MTS.

Frobisher Bay

RESOLUTION

Kimmirut

HUDSON BAY

All islands within bays and straits lie within Nunavut.

NOTTINGHAM ISLAND

SALISBURY

Hudson Strait

C. DE NOUVELLE-FRANCE

AKPATOK

KILLINIQ I.

TORNGAT MTS.

Hebron

Nain

NEWFOUNDLAND

Hopedale
Makkovik

Rigolet

Hamilton Inlet

Cartwright

Battle Harbour

PENINSULE D'UNGAVA

Ungava Bay

Kuujjuaq

MEALY MTS.

Happy Valley-Goose Bay

LABRADOR

Churchill Falls

Strait of Belle Isle

St. Anthony

Povungnituk

OTTAWA ISLANDS

MTS. OTISH

Schefferville

Corner Brook
Stephenville
St. George

LONG RANGE MTS.

GROS MORNE NAT'L PARK

BELCHER ISLANDS

Ft. Severn

C. HENRIETTA MARIA

PTE. LOUIS-XIV

AKIMISKI I.

Chisasibi

James Bay

Ft. Albany

Moosonee

ONTARIO

Fraserdale

Coral Rapids

La Sarre

Amos

Rouyn

Val-d'Or

Senneterre

Parent

La Tuque

Sept-Îles

ÎLE D'ANTICOSTI

Natashquan

Mingan

Clarke City

QUEBEC

Baie-Comeau

Chibougamau

Dolbeau

Alma

St. Félicien

Roberval
Chambord

Jonquière
Chicoutimi

La Malbaie

Gulf of St. Lawrence

Gaspé

MTS. CHIC-CHOCS

Matane

New Carlisle

Chandler

Carleton

PEI

Armstrong Sta.

Nakina

Hearst

Kapuskasing

Cochrane

Iroquois Falls

Timmins

Kirkland Lake

Cobalt

Ville-Marie

Temiscaming

St. Lawrence River

Montmagny

Lévis

Québec

Rimouski

Trois-Pistoles

Rivière-du-Loup

Edmundston

Campbellton

Bathurst

Chatham

NEW BRUNSWICK

PRINCE EDWARD I.

PRINCE EDWARD NAT'L PARK

Summerside

Charlottetown

Amherst

New Waterford

Sydney

Glace Bay

Geraldton

Longlac

Obo

Timmins

Chapleau

North Bay

Mattawa

Pembroke

Renfrew

Sudbury

Sturgeon Falls

Espanola

Blind River

Sault Ste. Marie

MANITOULIN I.

Georgian Bay

Thessalon

Huntsville

Bancroft

Smiths Falls

Ottawa

Brockville

Kingston

Alexandria Bay

Ogdensburg

Sorel

St-Hyacinthe

Drummondville

Joliette

Hull

MONTREAL

Shawinigan

Trois-Rivières

Sherbrooke

Granby

Victoriaville

Thetford Mines

MAINE

Fredericton

Woodstock

St. Andrews

St. Stephen

Moncton

Springhill

FUNDY NAT'L PARK

Truro

Kentville

NOVA SCOTIA

Saint John

St. George

Digby

Annapolis

Dartmouth

Halifax

Lunenburg

Bridgewater

Liverpool

Shelburne

Yarmouth

CAPE SABLE

ATLANTIC OCEAN

CAPE COD

Thunder Bay

PUKASKWA NAT'L PARK

MICHIPICOTEN

Lake Superior

Marathon

Marquette

Escanaba

Sault Ste. Marie

M I C H I G A N

Wiarton

Owen Sound

Midland

Orillia

Barrie

Parry Sound

Georgian Bay

Collingwood

Peterborough

Trenton

Cobourg

Oshawa

Whitby

TORONTO

Hamilton

Kincardine

Kitchener

NEW YORK

Rochester

Cobourg

Lake Ontario

VERMONT

Montpelier

NEW HAMPSHIRE

Concord

Portland

Ogdensburg

Albany

Schenectady

MASS.

Hartford

CONN.

Providence

R.I.

BOSTON

Duluth

Superior

MINNESOTA

St. Paul

MINNEAPOLIS

Madison

WISCONSIN

Green Bay

MILWAUKEE

CHICAGO

ILL.

Lake Michigan

Grand Rapids

Lansing

Flint

Saginaw

DETROIT

Toledo

Windsor

Chatham

Sarnia

Port Huron

London

St. Thomas

Niagara Falls

St. Catharines

BUFFALO

N E W   Y O R K

Rochester

Scranton

PENNSYLVANIA

N.J.

Newark

Lake Erie

OHIO

Red Lake

Sioux Lookout

Dryden

Lake of the Woods

Rainy

Nipigon

Lake Nipigon

Thunder Bay

A-520200-76

COPYRIGHT BY

RAND McNALLY & COMPANY

MADE IN U.S.A.

Relief

Meters		Feet
3050		10 000
1525		5000
610		2000
305		1000
152.5		500
0	Sea Level	0
152.5		500
1525		5000
3050		10 000

---

0 25 50 75 100	200	300	400	500 Miles
0 100	200	400	600	800 Kilometers

Scale 1:12 000 000; one inch to 190 miles. Polyconic Projection
Elevations and depressions are given in feet

Cities
and
Towns

0 to 50,000   500,000 to 1,000,000

50,000 to 500,000   1,000,000 and over

Scale 1:12 000 000; one inch to 190 miles. Polyconic Projection
Elevations and depressions are given in feet

Relief		
Meters		Feet
3050		10 000
1525		5000
610		2000
305		1000
152.5		500
Sea Level		0
		Below
152.5		Sea Level
		500
1525		5 000
3050		10 000
6100		20 000

Cities and Towns

| 0 to 50,000 | o | 500,000 to 1,000,000 | ◎ |
| 50,000 to 500,000 | ⊙ | 1,000,000 and over | |

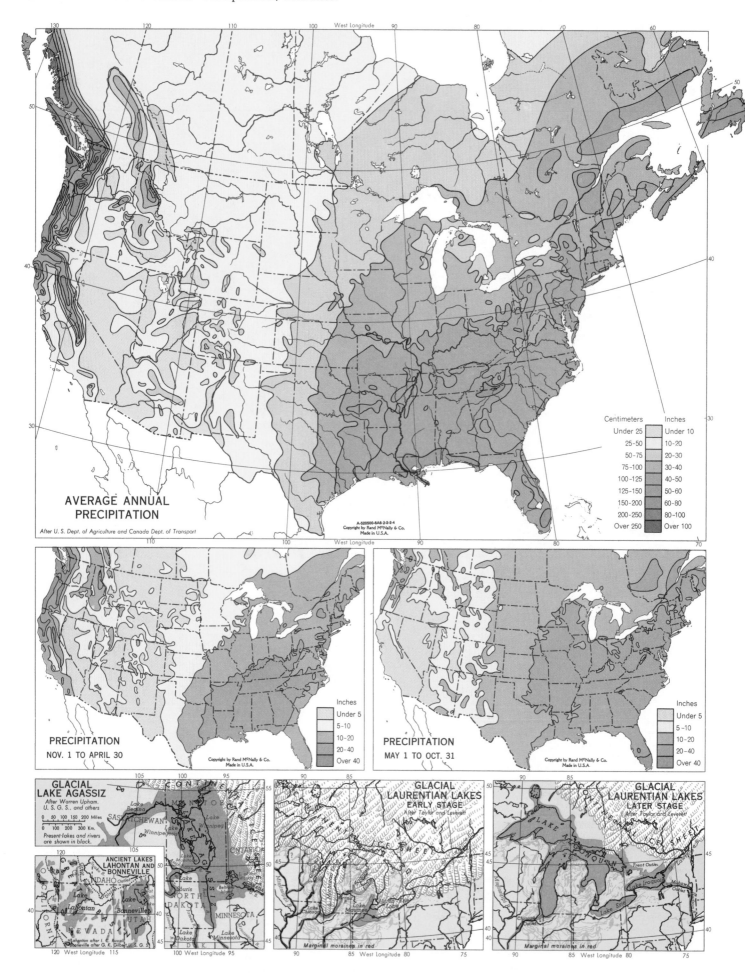

AVERAGE ANNUAL PRECIPITATION

After U.S. Dept. of Agriculture and Canada Dept. of Transport

A-520500-6A8-2-2-2-4
Copyright by Rand McNally & Co.
Made in U.S.A.

Centimeters	Inches
Under 25	Under 10
25-50	10-20
50-75	20-30
75-100	30-40
100-125	40-50
125-150	50-60
150-200	60-80
200-250	80-100
Over 250	Over 100

PRECIPITATION
NOV. 1 TO APRIL 30

Copyright by Rand McNally & Co.
Made in U.S.A.

Inches
Under 5
5-10
10-20
20-40
Over 40

PRECIPITATION
MAY 1 TO OCT. 31

Copyright by Rand McNally & Co.
Made in U.S.A.

Inches
Under 5
5-10
10-20
20-40
Over 40

GLACIAL LAKE AGASSIZ
After Warren Upham,
U. S. G. S., and others

0 50 100 150 200 Miles
0 100 200 300 Km.

Present lakes and rivers
are shown in black.

ANCIENT LAKES LAHONTAN AND BONNEVILLE
Lahontan after I. E. Russell
Bonneville after G. K. Gilbert, U. S. G. S.

GLACIAL LAURENTIAN LAKES
EARLY STAGE
After Taylor and Leverett

Marginal moraines in red

GLACIAL LAURENTIAN LAKES
LATER STAGE
After Taylor and Leverett

Marginal moraines in red

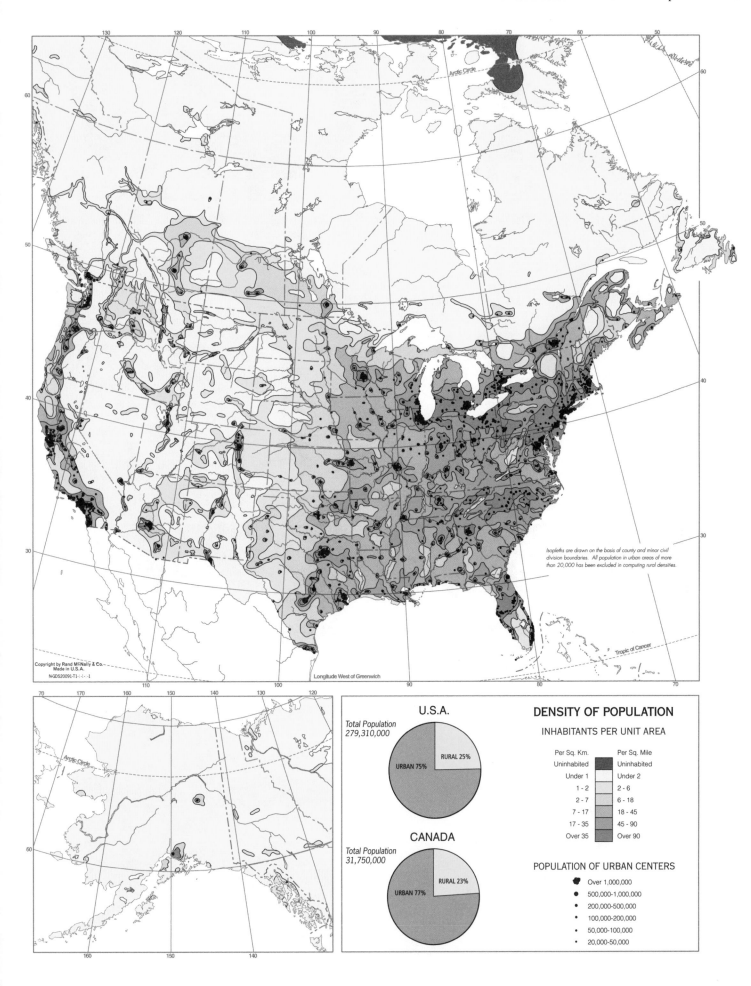

Isopleths are drawn on the basis of county and minor civil division boundaries. All population in urban areas of more than 20,000 has been excluded in computing rural densities.

Copyright by Rand McNally & Co.
Made in U.S.A.
N-GDS20091-T1

Longitude West of Greenwich

U.S.A.

Total Population
279,310,000

URBAN 75%   RURAL 25%

CANADA

Total Population
31,750,000

URBAN 77%   RURAL 23%

DENSITY OF POPULATION

INHABITANTS PER UNIT AREA

Per Sq. Km.	Per Sq. Mile
Uninhabited	Uninhabited
Under 1	Under 2
1 - 2	2 - 6
2 - 7	6 - 18
7 - 17	18 - 45
17 - 35	45 - 90
Over 35	Over 90

POPULATION OF URBAN CENTERS

- Over 1,000,000
- 500,000-1,000,000
- 200,000-500,000
- 100,000-200,000
- 50,000-100,000
- 20,000-50,000

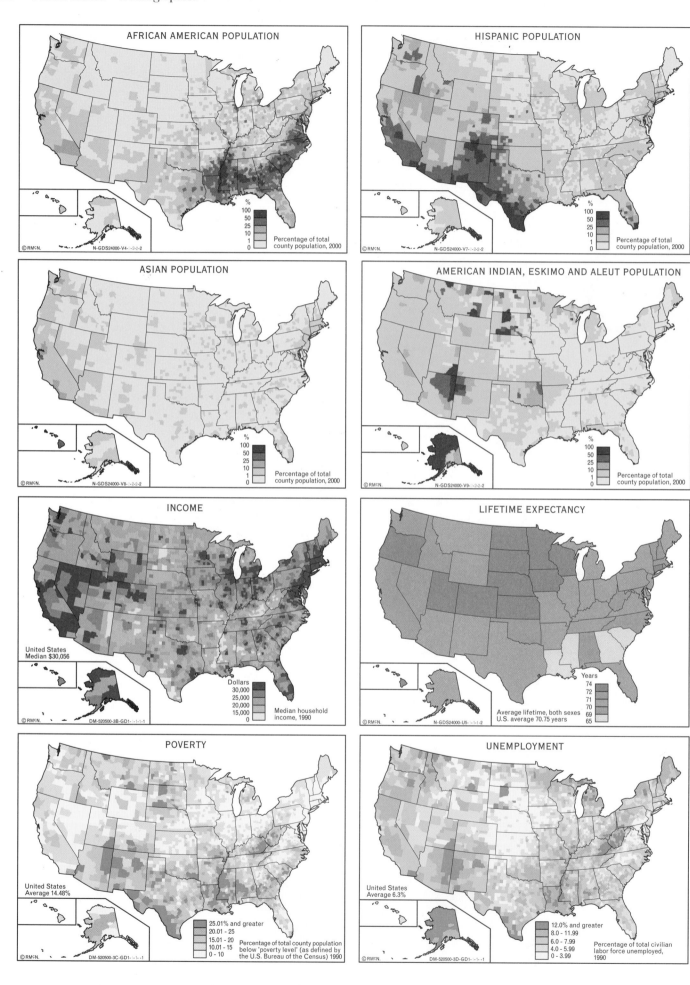

AFRICAN AMERICAN POPULATION

%
100
50
25
10
1
0

Percentage of total county population, 2000

N-GDS24000-V4--2-2-2

HISPANIC POPULATION

%
100
50
25
10
1
0

Percentage of total county population, 2000

N-GDS24000-V7--2-2-2

ASIAN POPULATION

%
100
50
25
10
1
0

Percentage of total county population, 2000

N-GDS24000-V8--2-2-2

AMERICAN INDIAN, ESKIMO AND ALEUT POPULATION

%
100
50
25
10
1
0

Percentage of total county population, 2000

N-GDS24000-V9--2-2-2

INCOME

United States Median $30,056

Dollars
30,000
25,000
20,000
15,000
0

Median household income, 1990

DM-520500-3B-GD1--1-1-1

LIFETIME EXPECTANCY

Years
74
72
71
70
69
65

Average lifetime, both sexes
U.S. average 70.75 years

N-GDS24000-U5--1-1-2

POVERTY

United States Average 14.48%

25.01% and greater
20.01 - 25
15.01 - 20
10.01 - 15
0 - 10

Percentage of total county population below 'poverty level' (as defined by the U.S. Bureau of the Census) 1990

DM-520500-3C-GD1--1-1-1

UNEMPLOYMENT

United States Average 6.3%

12.0% and greater
8.0 - 11.99
6.0 - 7.99
4.0 - 5.99
0 - 3.99

Percentage of total civilian labor force unemployed, 1990

DM-520500-3D-GD1--1-1-1

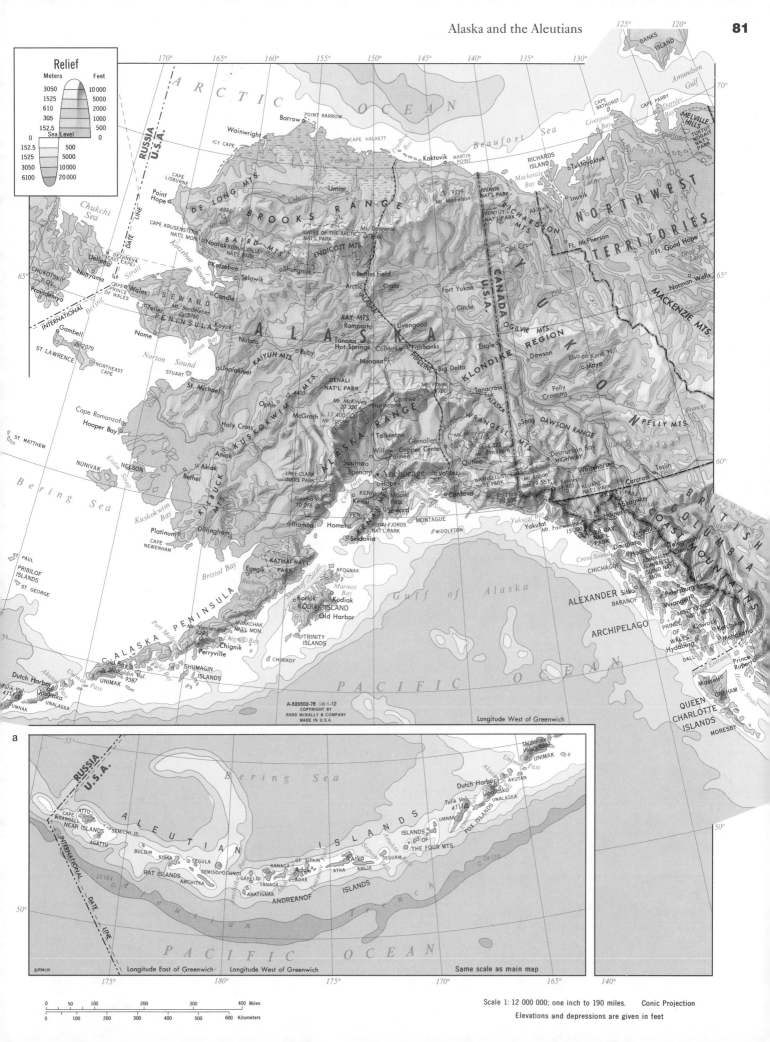

Relief

Meters	Feet
3050	10 000
1525	5000
610	2000
305	1000
152.5	500
0 Sea Level	0
152.5	500
1525	5000
3050	10 000
6100	20 000

A-520502-76 9-6-6-12
COPYRIGHT BY
RAND McNALLY & COMPANY
MADE IN U.S.A.

Longitude West of Greenwich

a

©RMcN. Longitude East of Greenwich   Longitude West of Greenwich   Same scale as main map

Scale 1: 12 000 000; one inch to 190 miles.   Conic Projection

Elevations and depressions are given in feet

| 0 | 50 | 100 | 200 | 300 | 400 Miles |
| 0 | 100 | 200 | 300 | 400 | 500 | 600 Kilometers |

BRITISH COLUMBIA

CANADA
U.S.A.

VANCOUVER ISLAND

Nanaimo
N. Vancouver
Vancouver
New Westminster
Ladysmith
Steveston
Duncan
Blaine
Lynden
Chilliwack
Grand Forks
Rossland
Trail
Strait of Georgia
Esquimalt
Victoria
Port Angeles
Port Townsend
Bellingham
SAN JUAN ISLANDS
Anacortes
Sedro Woolley
Concrete
Newhalem
Mt. Baker 10,778
Oroville
Northport
Porthill
Bonners Ferry
CAPE FLATTERY
MAKAH IND. RES.
Strait of Juan de Fuca
Mount Vernon
Arlington
NORTH CASCADES NAT'L PARK
Okanogan
COLVILLE IND. RES.
Colville
Chewelah
Sandpoint
Troy
Libby
CABINET MTS.

OLYMPIC MTS.
OLYMPIC NATIONAL PARK
Mt. Olympus 7965
Snohomish
Everett
Glacier Peak 10,541
Chelan
WELLS DAM
GRAND COULEE DAM
Mansfield
SPOKANE IND. RES.
Deer Park
Newport
Spirit Lake
Lake Pend Oreille
KALISPEL IND. RES.

Seattle
Bremerton
Kirkland
Bellevue
Renton
Monroe
Leavenworth
Cashmere
Cascade Tunnel
Waterville
Davenport
Spokane
Medical Lake
Cheney
Opportunity
Coeur d'Alene
Kellogg
Wallace
Mullan
Thompson Falls

QUINAULT IND. RES.
Moclips
Shelton
Tacoma
Lakewood Center
Auburn
Puyallup
Enumclaw
Roslyn
Cle Elum
WENATCHEE MTS.
ROCK ISLAND DAM
Ephrata
Odessa
COEUR D'ALENE IND. RES.
St. Maries

WASHINGTON

Hoquiam
Aberdeen
Montesano
Elma
Olympia
Carbonado
Mt. Rainier 14,410
Ellensburg
Moses Lake
Ritzville
Colfax
Palouse
Moscow
Elk River

Grays Harbor
Cosmopolis
Raymond
South Bend
Centralia
Chehalis
MOUNT RAINIER NATIONAL PARK
Yakima
PRIEST RAPIDS DAM
LOWER MONUMENTAL DAM
Pullman
LOWER GRANITE DAM

Willapa Bay
Castle Rock
Longview
Kelso
Kalama
Mt. Saint Helens 8364
Toppenish
Sunnyside
Richland
Pasco
Waitsburg
Dayton
Pomeroy
LITTLE GOOSE DAM
Clarkston
Lewiston
NEZ PERCE IND. RES.
Winchester

Ilwaco
Warrenton
Astoria
Rainier
Saint Helens
YAKIMA INDIAN RESERVATION
Mt. Adams 12,276
Prosser
Kennewick
Wallula
ICE HARBOR DAM
Walla Walla
Milton-Freewater
McNary DAM
Asotin
Nez Perce

Columbia
Seaside
Mt. St. Helens

Tillamook Bay
Hillsboro
Forest Grove
Tillamook
Portland
Camas
Vancouver
Gresham
Hood River
BONNEVILLE DAM
THE DALLES DAM
The Dalles
Wasco
JOHN DAY DAM
Goldendale
Pendleton
Elgin
Wallowa
Enterprise
Grangeville
CLEARWATER MOUNTAINS

Milwaukie
Lake Oswego
Oregon City
W. Linn
Newberg
McMinnville
Sheridan
Dallas
Woodburn
Silverton
Mt. Hood 11,239
Heppner
Condon
La Grande
Union
WALLOWA MTS.
HELLS CANYON
New Meadows

Salem
Independence
Albany
Corvallis
Lebanon
WARM SPRINGS IND. RES.
Mt. Jefferson 10,497
Lake Simtustus
Lake Billy Chinook
John Day
Baker
BLUE MOUNTAINS

Newport
Toledo
Green Peter Lake
OREGON
Prineville
Bend
Prineville Res.
Crooked
Weiser
Payette
Ontario
Vale
IDAHO
SALMON RIVER

Eugene
Springfield
McKenzie
Cougar Res.
Lookout Pt. Lake
Reedsport
Cottage Grove
Hills Creek Lake
Crane Prairie Reservoir
GREAT SANDY DESERT
Burns
Warm Sprs. Res.
Beulah Res.
Emmett
Caldwell
Boise
Nampa

North Bend
Coos Bay
Coquille
Roseburg
Diamond Peak 8744
HARNEY BASIN
Malheur Lake
Lake Owyhee
Mountain Home
Glenns Ferry

CAPE BLANCO
Bandon
Myrtle Point
CRATER LAKE NATIONAL PARK
Mt. Scott 8926
Lake Sumner
Harney Lake
OWYHEE MTS.

Grants Pass
Lake Abert
STEENS MTN.
Jordan Cr.

Medford
Mt. McLoughlin 9495
KLAMATH MTS.
Klamath Falls
NEVADA

Brookings
Ashland
OREGON CAVES NAT'L MON.
CASCADE-SISKIYOU NAT'L MON.
Lakeview
FORT McDERMITT IND. RES.
DUCK VALLEY IND. RES.

Crescent City
Happy Camp
Yreka
Lower Klamath Lake
Clear Lake Res.
WARNER MTS.
Upper Lake
PINE FOREST RA.
SANTA ROSA RA.
INDEPENDENCE MTS.

REDWOOD N.P.
HOOPA VALLEY IND. RES.
Weed
Mt. Shasta 14,162
LAVA BEDS NAT'L MON.
Alturas
Lower Lake
SUMMIT LAKE IND. RES.
Paradise Valley
Midas
Tuscarora

Arcata
Fieldbrook
CALIFORNIA
Mt. Shasta
Dunsmuir
Eagle Peak 9892
BLACK ROCK DESERT
Winnemucca
Wells

Humboldt Bay
Eureka
Fortuna
Ferndale
CAPE MENDOCINO
Scotia
Weaverville
Redding
Anderson
LASSEN VOLCANIC NATIONAL PARK
Lassen Peak (Vol.) 10,457
Eagle Lake
SMOKE CREEK DESERT
Battle Mountain
Elko

PACIFIC OCEAN

COAST RANGE

CASCADE RANGE

C A S C A D E   R A N G E

Longitude West of Greenwich

Scale 1: 4,000,000; one inch to 64 miles. Conic Projection
Elevations and depressions are given in feet

A-520597-76
COPYRIGHT BY
RAND McNALLY & COMPANY
MADE IN U.S.A.

ALBERTA SASKATCHEWAN

CANADA
U.S.A.

WATERTON-GLACIER INTERNATIONAL PEACE PARK

BLACKFEET IND. RES.

Sunburst Morgan Plentywood
Hogeland Opheim Scobey Grenora

Cut Bank Chinook Harlem Malta FORT PECK IND. RES.
Shelby Havre Ft. Belknap Glasgow Wolf Point Poplar Williston

Browning Lake Elwell FT. BELKNAP IND. RES. Medicine Lake
Whitefish Valier Conrad ROCKY BOYS IND. RES. Ft. Peck Sidney 48°
Kalispell Choteau Fort Benton Winifred Fort Peck Lake N. DAK.

Ronan NATIONAL BISON RANGE Missouri Brockway Glendive Beach

Missoula Great Falls Belt Lewistown Winnett Terry
Lolo Helena LITTLE BELT MTS. Harlowton Roundup Miles City Baker Marmarth
Stevensville East Helena Neihart White Sulphur Spgs. Musselshell Forsyth
Hamilton Deer Lodge Townsend Harlowton

Philipsburg BIG BELT MTS. CRAZY MTS. Colstrip
Anaconda Walkerville Three Forks Big Timber Huntley Billings Hardin Lame Deer
Butte Bozeman Livingston Columbus Laurel Crow Agency NORTHERN CHEYENNE IND. RES. Boxelder

BIG HOLE NAT'L BATTLEFIELD Twin Bridges Red Lodge CROW IND. RES. LITTLE BIGHORN BATTLEFIELD NAT'L MON.
Homer Youngs Peak 10,621 PIONEER MTS. Granite Peak 12,799 Bear Creek Sheridan DEVILS TOWER NAT'L MON.
Salmon Dillon Electric Peak 10,992 Gardiner Bighorn Lake BIGHORN MOUNTAINS Syndaces

Mt. Washburn 10,243 Lovell Powell Cloud Peak 13,167 Buffalo Gillette Moorcroft
YELLOWSTONE NATIONAL PARK Cody Greybull Basin Ten Sleep 44°
7733 ft above sea level Shoshone Lake Worland Kaycee
Borah Pk. 12,662 LOST RIVER RA. St. Anthony Ashton Gebo Midwest
LEMHI RANGE Mackay Rexburg GRAND TETON NAT'L PARK Thermopolis
Hyndman Peak 12,009 Arco Rigby Grand Teton 13,770 WIND RIVER IND. RES.
CRATERS OF THE MOON NAT'L MON. Idaho Falls Shelley Gannett Peak 13,804 Shoshoni Powder River
SNAKE RIVER PLAIN FORT HALL Fremont Peak 13,745 Riverton Glenrock
Shoshone Pocatello IND. RES. Lander Casper Douglas Orin
American Falls Soda Springs WIND RIVER RANGE WYOMING
Rupert Burley Lava Hot Sprs. Meade Peak 9957 Pathfinder Res. Alcova Res.
Twin Falls Oakley Montpelier GREAT DIVIDE BASIN Seminoe Res. Hanna Wheatland
Malad City Preston Kemmerer Sweetwater Rawlins Medicine Bow
Lewiston Richmond Superior 42°
Smithfield Logan Granger Green River Rock Springs
Providence Flaming Gorge Res.
Wellsville Brigham PARK RANGE
GREAT SALT LAKE Huntsville Evanston
GREAT SALT LAKE DESERT Ogden Morgan UINTA MTS. DINOSAUR NAT'L MON.
Wendover Farmington Kings Peak 13,498 Vernal
Bountiful Salt Lake City Park City UINTAH AND OURAY IND. RES. Craig
UTAH Murray Midvale COLO.
Tooele Heber City Steamboat Spgs.
Oak Creek

0   20   40   60   80   100   120 Miles
0   20   40   60   80   100   120   140   160   180   200 Kilometers

Relief

Meters		Feet
3050		10000
1525		5000
610		2000
305		1000
152.5		500
0	Sea Level	0
1525		500

Scale 1:4 000 000; one inch to 64 miles. Conic Projection
Elevations and depressions are given in feet

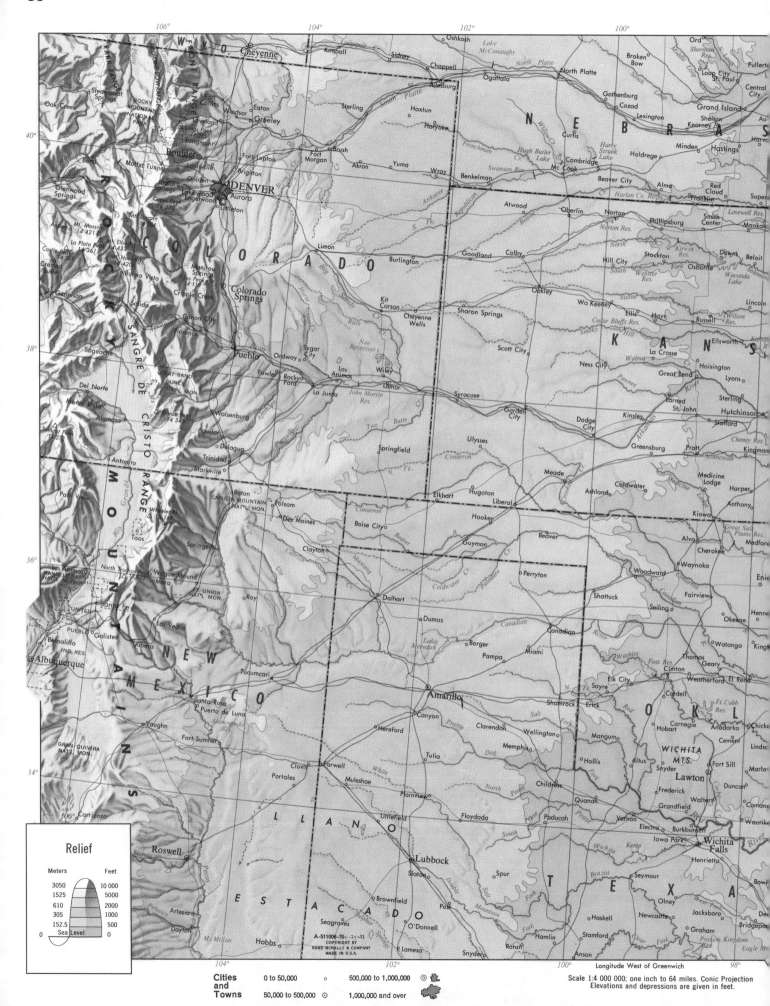

86

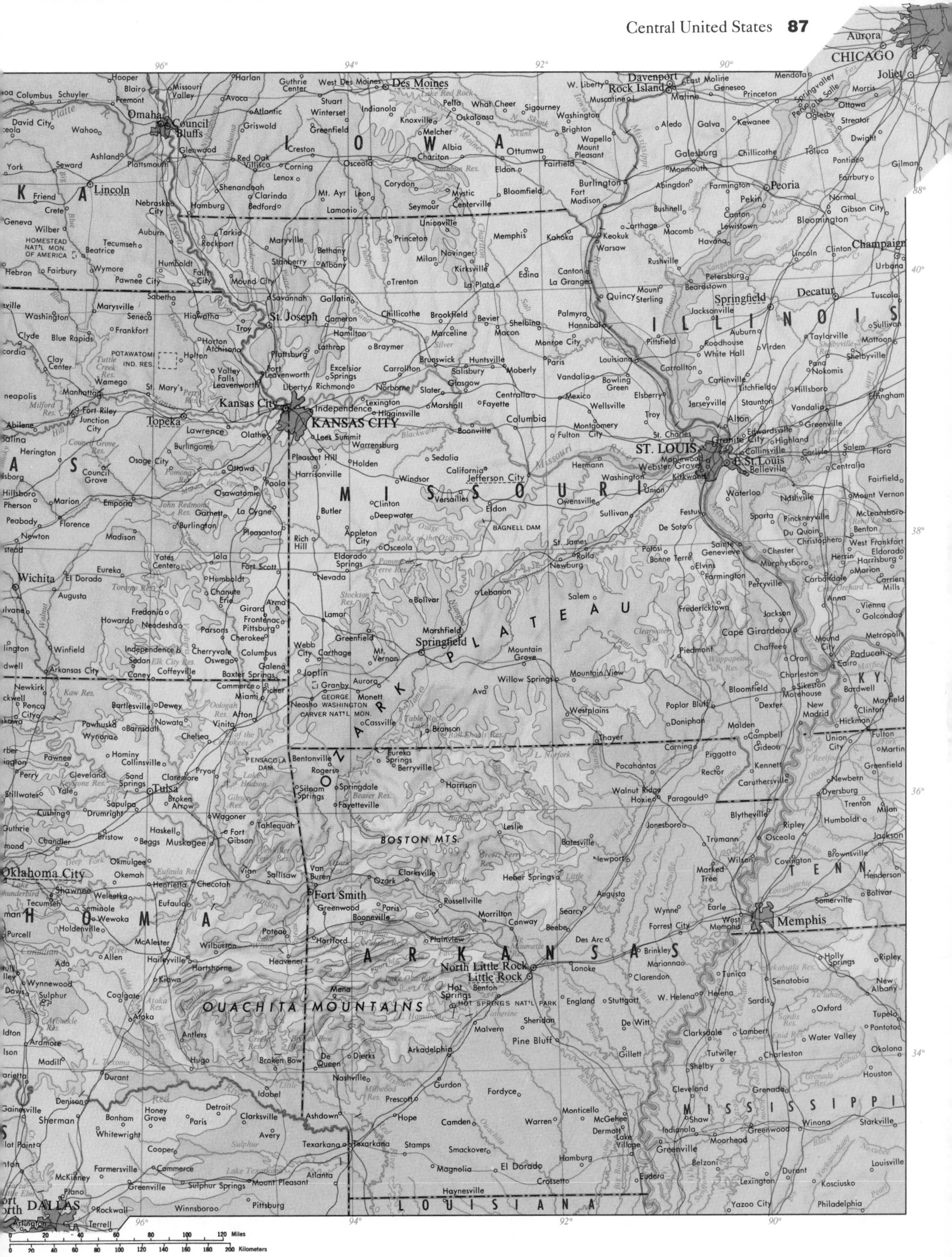

IOWA

KANSAS

MISSOURI

ILLINOIS

OKLAHOMA

ARKANSAS

TENN.

KY.

MISSISSIPPI

LOUISIANA

OZARK PLATEAU

BOSTON MTS.

OUACHITA MOUNTAINS

Lake of the Ozarks

CHICAGO

Aurora Joliet

Omaha Council Bluffs

Des Moines

Lincoln

St. Joseph

KANSAS CITY

Topeka

ST. LOUIS

St. Louis

Springfield

Wichita

Tulsa

Oklahoma City

Fort Smith

Little Rock

North Little Rock

Hot Springs

Memphis

West Memphis

DALLAS

Peoria

Bloomington

Champaign

Decatur

Paducah

Cape Girardeau

Jefferson City

Columbia

Quincy

Bagnell Dam

Pensacola Dam

George Washington Carver Nat'l Mon.

Homestead Nat'l Mon. of America

Hot Springs Nat'l Park

Potawatomi Ind. Res.

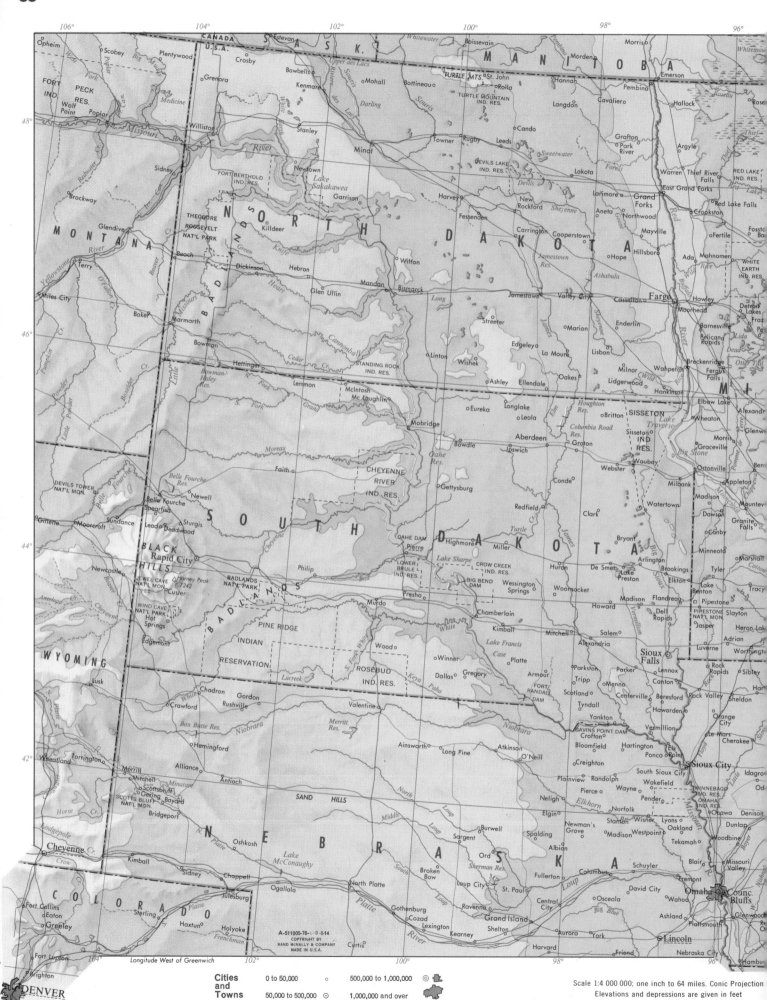

Cities and Towns

Scale 1:4 000 000; one inch to 64 miles. Conic Projection
Elevations and depressions are given in feet

Relief

Meters		Feet
1525		5000
610		2000
305		1000
152.5		500
0	Sea Level	0
152.5		500

Cities and Towns

0 to 50,000   500,000 to 1,000,000

50,000 to 500,000   1,000,000 and over

Longitude West of Greenwich

Scale 1:4 000 000; one inch to 64 miles.   Conic Projection
Elevations and depressions are given in feet

78° 76° 74° 72° 70°

QUEBEC

ALGONQUIN
PROVINCIAL
PARK

Opeongo
Bernard

Pembroke
Renfrew
Arnprior
Almonte
Aylmer
Hull Ottawa

St. Jerome
Terrebonne
Grenville Ste. Therese
Hawkesbury St. Jean
Vaudreuil Lambert
Beauharnois
Valleyfield Iberville
Huntingdon Bedford
Farnham

Richmond
East
Anguso
Scotstown

MONTREAL

St. Hyacinthe
Bromptonville
Granby Sherbrooke
Waterloo Magog

MAINE

Flagsta

Rimouski

CANADA
U.S.A.

Cornwall
Massena
Malone
Dannemora
Plattsburgh
Ogdensburg
Potsdam

Newport
St. Albans
Winooski
Burlington
Essex
Montpelier

St. Johnsbury
Groveton
Berlin
Whitefield

MT. WASHINGTON
6288

Rumford
Mexico
Augusta
Winthrop
Norway
Lewiston Lisbon
Brunswick

Peterborough
Kingston
Belleville Napanee
Lyons
Trenton
Picton

PRINCE EDWARD
PEN.

THOUSAND
ISLANDS
WOLFE ISLANDS
AMHERST

PETER PT.

Alexandria Bay Gouverneur
Carthage
Watertown
Lowville
Sackets Harbor

Saranac Lake
Tupper Lake
Lake Placid

Crown Point
Ticonderoga

MT. MANSFIELD
4393

Middlebury
Randolph
Proctor
Rutland
Wallingford
Brandon

Hanover
Lebanon
Meredith Laconia
Woodsville

MTS.
4310

VERMONT

NEW
HAMPSHIRE

Conway

Westbrook
Portland
Biddeford

Sanford North Berwick
Rochester Dover Kittery
Franklin
Concord Portsmouth

LAKE ONTARIO
Surface 245 Feet above Sea Level
maximum depth 802 Feet

TORONTO

Oshawa
Whitby Bowmanville
Newmarket
Orangeville
Brampton

Hamilton
St. Catharines
Brantford
Dunnville Welland
Thorold
Port Dalhousie

BUFFALO
Hamburg
Lackawanna
Silver Creek
Dunkirk
Fredonia
Mayville
Westfield

LONG PT.

Niagara Falls
Lockport Medina Albion
N. Tonawanda Brockport
Tonawanda
Lancaster Batavia
Attica
Warsaw
Perry

Oswego
Fulton

Rochester
E. Rochester
Newark
Baldwinsville
Seneca Falls Savoy
Waterloo
Geneva Auburn
Canandaigua
Mt. Morris
Penn Yan Cortland

NEW   YORK

Rome
Oneida Utica
Canastota
Syracuse
Skaneateles Cazenovia
Cooperstown
Norwich
Oneonta

Little Falls
Johnstown
Amsterdam
Gloversville
Dolgeville

Ballston Spa
Saratoga Springs
Schenectady
Troy
ALBANY

CATSKILL
MTS.

ADIRONDACK
MTS.

GREEN MTS.

Glens Falls
Whitehall
Granville

Bennington
North Adams
Greenfield
Pittsfield
Northampton
Holyoke Chicopee
Westfield Springfield

Fitchburg Lowell
Gardner
Leominster Clinton
Worcester
Webster

Lawrence
Haverhill
Newburyport
Beverly
Salem
Lynn
Somerville
Cambridge
Quincy
Weymouth
Brockton

C. ANN

BOSTON

MASS.

CONN.

Winsted
Torrington
Hartford
Bristol New Britain
Waterbury Meriden
New Haven
Bridgeport Stratford
Norwalk
Stamford

Willimantic
Windham
Putnam
Norwich

Woonsocket
Pawtucket
PROVIDENCE
Cranston
Warwick
Newport

Fall River
New Bedford

R.I.

NARRAGANSETT
BAY

MARTHAS
VINEYARD

C. COD

LONG ISLAND SOUND

White Plains
Yonkers
Mt. Vernon
New Rochelle

Greenport
MONTAUK PT.
Montauk
Southampton

BLOCK

Binghamton
Elmira
Corning
Hornell
Bath
Wellsville
Olean
Salamanca
Jamestown
Warren

Ithaca
Watkins Glen
Owego
Endicott

Scranton
Wilkes Barre
Hazleton

POCONO MTS.

Poughkeepsie
Newburgh
Kingston
Middletown
Port Jervis
Peekskill

PENNSYLVANIA

PITTSBURGH
McKeesport
Monongahela
Connellsville
Uniontown

Johnstown
Altoona
Indiana
Greensburg

Williamsport
Lock Haven
Jersey Shore
Clearfield
Du Bois
Brookville
Ridgway
St. Marys
Emporium
Renovo

Harrisburg
Carlisle
York
Lancaster
Reading
Lebanon
Pottsville
Allentown
Bethlehem
Easton

Phillipsburg
Trenton
New Brunswick
Princeton

NEW
JERSEY

NEW YORK

Newark
Jersey City
Elizabeth
Bayonne

East Orange
Paterson
Clifton
Passaic

SANDY HOOK
Long Branch
Asbury Park
Ocean Grove
Freehold
Lakewood

ATLANTIC

OCEAN

Atlantic City
Ventnor
Ocean City

Cape May C.H.
Wildwood
Cape May

MARYLAND

BALTIMORE
Dundalk
Towson
Catonsville
Annapolis

Frederick
Westminster
Hagerstown
Cumberland
Hancock
Gettysburg
Chambersburg
Waynesboro

Wheaton
Bethesda
Silver Spr.
Arlington
Alexandria
WASHINGTON D.C.

MT. VERNON

GEO. WASHINGTON
BIRTHPLACE
NAT'L MON.

DEL.

Dover
Milford
Seaford
Georgetown
Salisbury
Snow Hill
Pocomoke City
Crisfield

CAPE HENLOPEN

Delaware
Bay

CHESAPEAKE BAY

VIRGINIA

Charlottesville
Lynchburg
Richmond
Hopewell
Williamsburg
Yorktown

C. CHARLES

Newport News Hampton
Portsmouth Norfolk
Virginia Beach

SHENANDOAH
NAT'L PARK

BLUE RIDGE

Relief

Meters		Feet
1525		5000
610		2000
305		1000
152.5		500
0	Sea Level	0
152.5		500
1525		5000
3050		10 000

A-520596-76
COPYRIGHT BY
RAND McNALLY & COMPANY
MADE IN U.S.A.

0 20 40 60 80 100 120 Miles
0 20 40 60 80 100 120 140 160 180 200 Kilometers

80° 78° 76° 74° 72°

44° 40° 38°

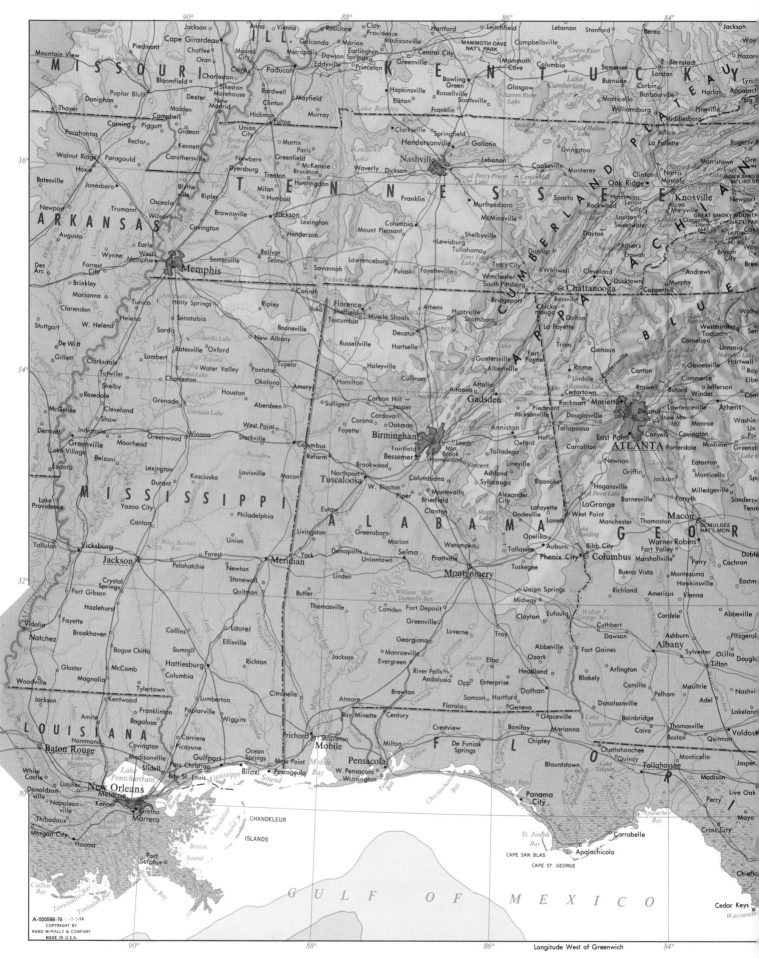

Scale 1:4 000 000; one inch to 64 miles. Conic Projection
Elevations and depressions are given in feet

Longitude West of Greenwich

GULF OF MEXICO

A-520598-76 -7-7-14
COPYRIGHT BY
RAND McNALLY & COMPANY
MADE IN U.S.A.

Relief

Meters	Feet
1525	5000
610	2000
305	1000
152.5	500
0	Sea Level	0
152.5	500
1525	5000

Same scale as main map

a

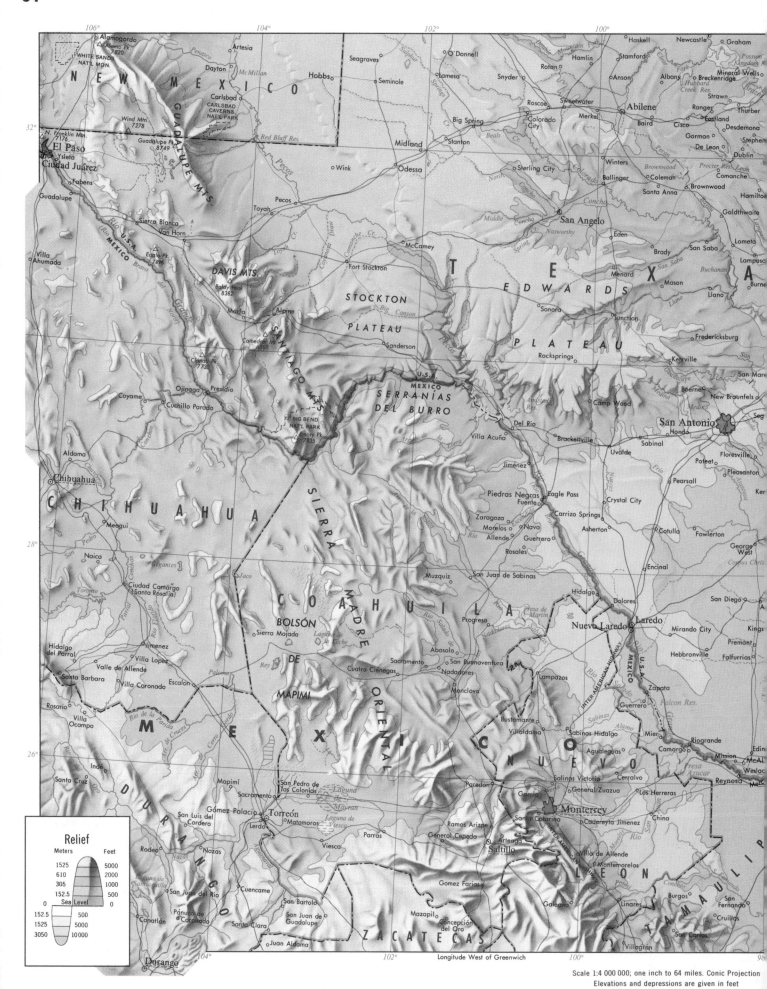

Relief

Meters		Feet
1525		5000
610		2000
305		1000
152.5		500
0	Sea Level	0
152.5		500
1525		5000
3050		10 000

Scale 1:4 000 000; one inch to 64 miles. Conic Projection
Elevations and depressions are given in feet

Longitude West of Greenwich

ARK.

MISSISSIPPI

LOUISIANA

Fort Worth DALLAS

HOUSTON

Corpus Christi

Brownsville
Matamoros

New Orleans

Baton Rouge

Shreveport

Monroe

Jackson

Vicksburg

Natchez

Alexandria

Lake Charles

Lafayette

Beaumont

Port Arthur

Galveston

Waco

Tyler

GULF OF MEXICO

Denton, McKinney, Farmersville, Greenville, Sulphur Springs, Mount Pleasant, Atlanta, Pittsburg, Winnsboro, Gilmer, Jefferson, Vivian, Haynesville, Homer, Bastrop, Lake Providence, Yazoo City, Canton, Forest, Ross Barnette Res., Pelahatchie

Plano, Rockwall, Terrell, Mineola, Marshall, Longview, Kilgore, Carthage, Bossier City, Minden, Arcadia, Rustono, Rayville, Delhi, Tallulah, Crystal Springs, Hazlehurst

Arlington, Waxahachie, Kaufman, Wills Point, Mabank, Henderson, Mansfield, Coushatta, Eros, Jonesboro, Winnsboro, Alto, Port Gibson, Fayette, Collins, Sumrall

Weatherford, Granbury, Cleburne, Itasca, Hillsboro, Hubbard, Corsicana, Athens, Jacksonville, Rusk, Timpson, Center, Natchitoches, Fisher, Colfax, Jonesville, Ferriday, Vidalia, Brookhaven, Norfield, McComb, Columbia, Tylertown, Lumberton, Poplarville

Meridian, Clifton, Mexia, Teague, Elkhart, Palestine, Nacogdoches, San Augustine, Hemphill, Peason, Pineville, Marksville, Woodville, Gloster, Magnolia, Franklinton

Waco, McGregor, Moody, Groesbeck, Buffalo, Ratcliff, Crockett, Lufkin, Leesville, Fullerton, Lecompte, McNary, Bunkie, Jackson, Kentwood, Amite, Covington, Picayune, Bay St. Louis

Temple, Cameron, Marlin, Marquez, Madisonville, Groveton, Jasper, Newton, De Ridder, Elizabeth, Oakdale, Ville Platte, Melville, New Roads, Hammond, Madisonville, Slidell

Bartlett, Hearne, Bryan, Huntsville, Trinity, Woodville, Wiergate, Longville, Kinder, Eunice, Opelousas, Plaquemine, White Castle, Donaldsonville, Lutcher, Kenner, Metairie, Gretna

Rockdale, Caldwell, Navasota, Willis, Conroe, Kirbyville, De Quincy, Jennings, Crowley, Rayne, St. Martinville, Napoleonville, Thibodaux

Taylor, Elgin, Giddings, Brenham, Hempstead, Cleveland, Saratoga, Silsbee, Vinton, Lake Arthur, Abbeville, New Iberia, Jeanerette, Franklin, Morgan City, Houma, Port Sulphur

Round Rock, Austin, Bastrop, Smithville, Lagrange, Bellville, Sealy, Humble, Baytown, Dayton, Liberty, Sourlake, Orange, Ged, Gueydan, Lake Pontchartrain

Lockhart, Luling, Gonzales, Columbus, Eagle Lake, Hallettsville, Wharton, Richmond, Alvin, Texas City, Port Bolivar, High Island

Cuero, Yoakum, El Campo, West Columbia, Angleton, Freeport, Bay City

Victoria, Goliad, Edna, Palacios, Port Lavaca, Matagorda

Refugio, Skidmore, Sinton, Rockport, Aransas Pass, St. Joseph, Mustang

Portland, Corpus Christi Bay, Padre, Island, Laguna Madre, Raymondville, Harlingen, San Benito

Toledo Bend Res., Sam Rayburn Res., Town Bluff Lake, Lake O' The Pines, Lake d'Arbonne, Lake Pontchartrain, Lake Borgne

Inset map (a):

HOUSTON

West University Place, Bellaire, Missouri City, Pearland, Arcola, Manvel, Sandy Point, Liverpool, Danbury, Angleton

Jacinto City, Galena Pk., Pasadena, South Houston, Genoa, Friendswood, Alvin, Algoa, Alta Loma, Hitchcock, La Marque, Hitchcock

Channelview, Baytown, La Porte, Seabrook, Kemah, League City, Dickinson, Texas City, Port Bolivar

Crosby, Sheldon, Highlands, Mont Belvieu, Wallisville, Hankamer, Anahuac

GALVESTON BAY, EAST BAY, Smith Point, High Island, BOLIVAR PENINSULA

GALVESTON ISLAND, Galveston

GULF OF MEXICO

Scale 1:1 000 000

0 5 10 Miles
0 4 8 12 16 Kilometers

Cities and Towns

| 0 to 50,000 | ○ | 500,000 to 1,000,000 | ◉ |
| 50,000 to 500,000 | ⊙ | 1,000,000 and over | |

0 20 40 60 80 100 120 Miles
0 20 40 60 80 100 120 140 160 180 200 Kilometers

Scale 1:16 000 000; one inch to 250 miles. Polyconic Projection
Elevations and depressions are given in feet

a

Caribbean Sea

PANAMA

Scale 1:1 000 000

Bahía de Panamá

A-530000-76 9 9-26 EL
COPYRIGHT BY
RAND MCNALLY & COMPANY
MADE IN U.S.A.

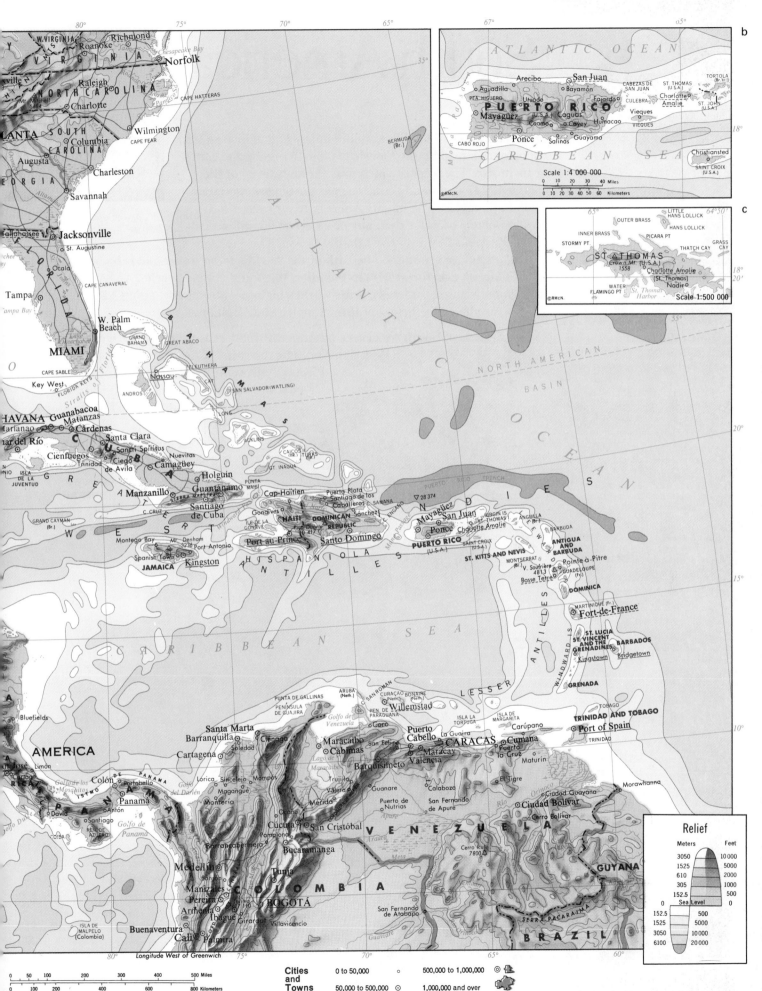

b

ATLANTIC OCEAN

Arecibo • San Juan
Aguadilla • Bayamón
PTA. HIGUERO • Utuado • CABEZAS DE SAN JUAN
Fajardo
ST. THOMAS
TORTOLA (Br.)
Charlotte Amalie
ST. JOHN (U.S.A.)

PUERTO RICO (U.S.A.)
Mayagüez • Caguas • Coamo • Covey • Humacao
Culebra
Vieques
VIEQUES

18°

CABO ROJO
Ponce • Salinas Guayama

MONA

CARIBBEAN SEA

Christiansted
SAINT CROIX (U.S.A.)

Scale 1:4 000 000
0 10 20 30 40 Miles
0 10 20 30 40 50 60 Kilometers
©RMCN

c

65°
LITTLE HANS LOLLICK
OUTER BRASS   HANS LOLLICK
64°50′
INNER BRASS   PICARA PT.
STORMY PT.   THATCH CAY   GRASS CAY
ST. THOMAS
Crown Mt. (U.S.A.)   18°
1558   20°
Charlotte Amalie (St. Thomas)
WATER   Nadir
FLAMINGO PT.   St. Thomas Harbor

Scale 1:500 000
©RMCN

W. VIRGINIA
ville
Roanoke   Richmond
VIRGINIA
Norfolk
Raleigh   Chesapeake Bay
Mt. Mitchell   NORTH CAROLINA
6684   Charlotte
CAPE HATTERAS
ANTA   SOUTH
Columbia   Wilmington
CAROLINA   CAPE FEAR
Augusta
GEORGIA   Charleston

35°

Savannah

Fallahassee   Jacksonville
St. Augustine
Ocala
Tampa   FLORIDA
Tampa Bay

BERMUDA (Br.)

ATLANTIC

MIAMI

W. Palm Beach
CAPE CANAVERAL

CAPE SABLE
Key West   Nassau
FLORIDA KEYS   GRAND BAHAMA
Straits of Florida   GREAT ABACO
ELEUTHERA
CAT
ANDROS   SAN SALVADOR (WATLING)
LONG

NORTH AMERICAN
BASIN

HAVANA   Guanabacoa
Marianao   Matanzas
Cárdenas
ar del Río   Santa Clara   ACKLINS
Cienfuegos   Sancti Spíritus   CROOKED
Trinidad   Ciego   Nuevitas
de Ávila   Camagüey   GT. INAGUA
ISLA   Holguín   CAICOS (BR.) TURKS
DE LA   PUNTA   OCEAN
JUVENTUD   MAISÍ
Manzanillo   Guantánamo   ▽ 28 374   PUERTO RICO TRENCH
GRAND CAYMAN   Santiago   Cap-Haïtien   Puerto Plata
(Br.)   de Cuba   Gonaïves   Santiago de los   SAMANA
Caballeros   Sánchez   Mayagüez   San Juan   VIRGIN IS.
Montego Bay   Mt. Denham   HAITI   DOMINICAN   Ponce   Charlotte Amalie   ST. THOMAS (Br.)   ANGUILLA
3236   Port Antonio   Pico Duarte   REPUBLIC   PUERTO RICO   ST. CROIX   BARBUDA
Spanish Town   ILE DE LA   3147   Santo Domingo   (U.S.A.)   ANTIGUA
JAMAICA   Port-au-Prince   HISPANIOLA   SAINT CROIX   AND BARBUDA
Kingston   GONAVE   ANTILLES   ST. KITTS AND NEVIS   Pointe-à-Pitre
MONTSERRAT (Br.)   GUADELOUPE (Fr.)
V. Soufrière   Basse-Terre
4813
CARIBBEAN   SEA   DOMINICA
MARTINIQUE (Fr.)
Fort-de-France

15°

LESSER   ST. LUCIA
ST. VINCENT AND THE   BARBADOS
GRENADINES   Bridgetown
WINDWARD IS.   Kingstown
ANTILLES   GRENADA

TOBAGO
ARUBA (Neth.)   SAN ROMAN   CURAÇAO BONAIRE (Neth.)   TRINIDAD AND TOBAGO
PUNTA DE GALLINAS   (Neth.)   Port of Spain
PENINSULA   PEN. DE   Willemstad   ISLA LA   TRINIDAD
DE GUAJIRA   PARAGUANA   TORTUGA   ISLA DE
Santa Marta   Coro   San Felipe   MARGARITA   Carúpano
Barranquilla   Ciénaga   Maracaibo   Puerto   La Guaira   CARACAS   Cumaná
Cartagena   Soledad   Cabello   Maracay   Puerto   10°
Lago de   Cabimas   Valencia   la Cruz
Maracaibo   Barquisimeto   Maturín
Bluefields   Trujillo   El Tigre   Morawhanna
AMERICA   Colón   PANAMA   Lorica   Sincelejo   Mompós   Guanare   Calabozo
José   Limón   Golfo del Darién   Valera   Ciudad Guayana
Portobello   Magangué   Mérida   Puerto de   San Fernando   Cerro Bolívar   Ciudad Bolívar
PANAMA   Montería   Nutrias   de Apure   7800
Panamá   Anton   VENEZUELA
David   Santiago   Barrancabermeja   Cúcuta   San Cristóbal   Apure   Cerro Icutú   GUYANA
PEN. DE   Bucaramanga   Pamplona   Arauca   7800   Río
AZUERO   Golfo de   San Fernando
Panamá   Medellín   Tunja   Meta   de Atabapo
Manizales   COLOMBIA   SERRA PACARAIMA
Pereira   BOGOTÁ   San Fernando
ISLA DE   Armenia   Ibagué   Girardot   de Atabapo   BRAZIL
MALPELO   Buenaventura   Villavicencio   Guaviare
(Colombia)   Cali   Palmira

80°   Longitude West of Greenwich   75°   70°   65°   60°

Relief

Meters	Feet
3050	10 000
1525	5000
610	2000
305	1000
152.5	500
0	Sea Level	0
152.5	500
1525	5000
3050	10 000
6100	20 000

0 50 100 200 300 400 500 Miles
0 100 200 400 600 800 Kilometers

Cities and Towns
0 to 50,000   ○   500,000 to 1,000,000
50,000 to 500,000   ⊙   1,000,000 and over

South America

Floodplain of the Amazon River, Brazil

With an area of 6.9 million square miles (17.8 million sq km), triangular-shaped South America is fourth among the continents in size. The Andes, which pass through seven of the continent's 13 mainland countries, are the longest mountain chain in the world. The mighty Amazon River carries a greater volume of water than any other river: 46 million gallons per second flow into the Atlantic Ocean. The Amazon basin contains an estimated one-fifth of the world's fresh water and is home to the world's largest rain forest with its countless plant and animal species. Angel Falls, in a remote Venezuelan forest, is the world's highest waterfall, dropping 3,212 feet (979 m), or almost the height of three Empire State Buildings.

One of South America's other great wonders is manmade. High in the Peruvian Andes lie the ruins of the sacred city of Machu Picchu, built centuries ago by the Incas. The city has an exquisite design and was built with remarkable skill. The Inca population, like most of South America's other native peoples, declined rapidly after the arrival of Europeans in the early 16th century.

South America at a glance

Land area: 6,900,000 square miles (17,800,000 sq km)

Estimated population: 352,960,000

Population density: 51/square mile (20/sq km)

Mean elevation: 1,800 feet (550 m)

Highest point: Aconcagua, Argentina, 22,831 feet (6,959 m)

Lowest point: Salinas Chicas, Argentina, 138 feet (42 m) below sea level

Longest river: Amazon-Ucayali, 4,000 mi (6,400 km)

Number of countries (incl. dependencies): 15

Largest independent country: Brazil, 3,300,172 square miles (8,547,404 sq km)

Smallest independent country: Suriname, 63,251 square miles (163,820 sq km)

Most populous independent country: Brazil, 175,260,000

Least populous independent country: Suriname, 435,000

Largest city: São Paulo, pop. 9,713,692

Wettest place:
Quibdó, Colombia
354 inches (899 cm)/year

Orinoco

Llanos

Guiana Highlands

Equator

Amazon

Amazon Basin

Madeira

São Francisco

Pacific Ocean

Lago Titicaca

Mato Grosso

Brazilian Highlands

Atlantic Ocean

Driest place:
Arica, Chile
.03 inches (.08 cm)/year

Gran Chaco

Paraná

Tropic of Capricorn

Hottest place:
Rivadavia, Argentina
120°F (49°C)

Paraguay

Highest point:
Cerro Aconcagua, Argentina
22,831 ft (6,959 m)

Pampas

Lowest point:
Salinas Chicas, Argentina
138 ft (42 m) below sea level

Coldest place:
Sarmiento, Argentina
-27°F (-33°C)

Patagonia

FALKLAND ISLANDS

TIERRA DEL FUEGO

Landforms

▨	Mountains
▨	Widely spaced mountains
▨	High tablelands
▨	Hills and low tablelands
▨	Plains
▨	Depresssions, basins
▨	High tablelands and ice caps
▨	Mountains and ice caps

© Rand McNally & Co.
M-540000-7C-EL1-˙-!- -1

The Andes, at the western edge of Argentina's Patagonia region.

© Rand McNally & Co.
M-540000-6A-EL1-'-'- -1'

Climate

South America's most predominant climate zones are the vast tropical rain forests and tropical savannas which cover most of the northern half of the continent. In the rain forests, rain falls throughout the year, averaging 60 to 80 inches (152 to 203 cm) annually. Daytime temperatures usually exceed 80° F (27° C). The tropical savanna regions experience the same high temperatures but less rainfall, with a dry season in winter. A temperate climate, with milder temperatures and moderate rainfall, prevails throughout much of southern South America, east of the Andes. Arid to semiarid conditions are found in the far south and at Brazil's eastern tip.

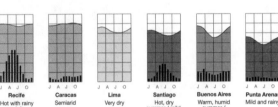

Tinted areas show temperature in degrees Fahrenheit. Vertical bars show precipitation in inches.

| **Manaus** Hot and rainy | **Recife** Hot with rainy and dry seasons | **Caracas** Semiarid | **Lima** Very dry | **Santiago** Hot, dry summer / mild, rainy winter | **Buenos Aires** Warm, humid summer / mild winter | **Punta Arenas** Mild and rainy | **Extensive uplands** Climate varies with elevation and latitude |

Population

South America shares the rank of fourth most-densely populated continent with North America, with 51 people per square mile (20 per sq km). Despite this relatively low figure, the continent is intensely urban because the Andes and the Amazon rain forest render most of it either inaccessible or unsuitable for farming. More than 90% of South America's 353 million people live within 150 miles (240 km) of the coast. São Paulo, Brazil, with a metropolitan population of more than 17 million, is the world's sixth-largest metropolitan area. Most South Americans are *mestizo*—of mixed European and Indian descent. Spanish is the predominant language, followed by Portuguese. More than 90% of the people are Roman Catholics.

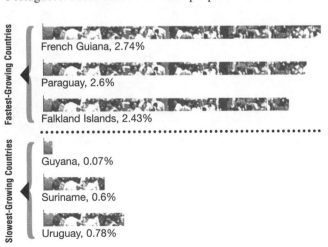

Fastest-Growing Countries

French Guiana, 2.74%

Paraguay, 2.6%

Falkland Islands, 2.43%

Slowest-Growing Countries

Guyana, 0.07%

Suriname, 0.6%

Uruguay, 0.78%

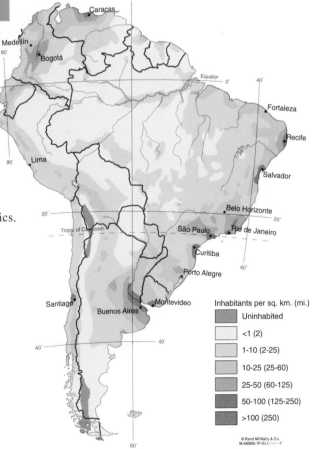

Inhabitants per sq. km. (mi.)

	Uninhabited
	<1 (2)
	1-10 (2-25)
	10-25 (25-60)
	25-50 (60-125)
	50-100 (125-250)
	>100 (250)

© Rand McNally & Co.
M-540000-1P-EL1-'-'- -1'

Environments and Land Use

Land suitable for farming is very limited in South America, covering only about 6% of the continent. Small, family-run subsistence farms are common, and typical crops are maize, wheat, and potatoes. Despite the scarcity of arable land, commercial agriculture for export is a major part of the economies of several countries. Ecuador is the world's leading exporter of bananas, while Brazil and Colombia grow almost 40% of the world's coffee beans. Brazil is also a major exporter of sugar. Chile has developed a large trade in produce—such as tomatoes and grapes—that is exported to North America during its winter months (South America's summer months). Production of coca, the basis for illicit drugs, has become a part of the rural economies of Colombia, Bolivia, and Peru.

Cattle ranching is centered on the vast, grassy Pampas region, which extends through northern Argentina, Uruguay, and southern Brazil. Sheep, raised both for meat and wool, are important throughout the Andes and southern Argentina. About 25% of the continent is suitable for grazing.

As South America's population grows, pressure builds to clear more land for farming. Much expansion has taken place in the Amazon basin at the cost of millions of acres of rain forest, which are cleared of trees and drained. Balancing the demands of the population with the need to preserve the rain forest is one of the continent's most pressing issues.

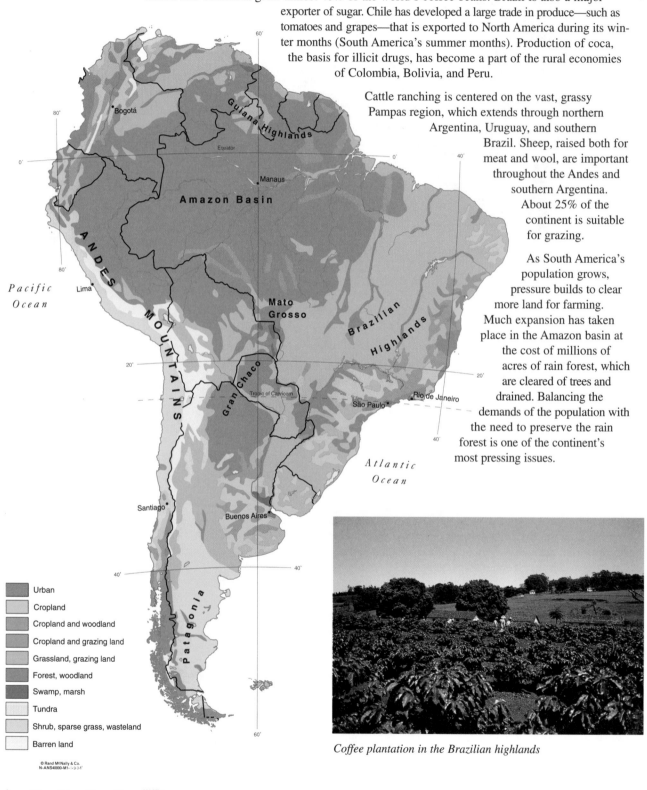

Urban

Cropland

Cropland and woodland

Cropland and grazing land

Grassland, grazing land

Forest, woodland

Swamp, marsh

Tundra

Shrub, sparse grass, wasteland

Barren land

© Rand McNally & Co.
N-ANS40000-M1-·-·-·-1'

0	200	400	600	800	1000 Miles

0	300	600	900	1200	1500 Kilometers

Coffee plantation in the Brazilian highlands

Destruction of the Rain Forest

The Amazonian rain forest contains an abundance and diversity of life that is matched by few places in the world. In fact, it has been estimated that the plant and animal species of Amazonia account for nearly one-half of those found on Earth. New plants and animals are constantly being discovered, and scientists have found in Amazonian plants a treasure trove of new substances, some of which are now being used to produce life-saving medicines. It is thought that cures for many more diseases could be found in the plants yet to be studied.

In recent decades, nearly 10% of the rain forest's original 1.58 million square miles (4.09 million sq km) has been cleared (see map below) for farming, cattle ranching, mining, and commercial logging. The most effective way to clear the land is the "slash-and-burn" method, which has been practiced by indigenous peoples on an insignificant scale for centuries. Today, widespread usage of this method is destroying vast areas of the rain forest, and smoke from the fires is polluting the atmosphere and possibly contributing to global warming. The destruction also imperils the Indians living within the forest; their numbers have shrunk by more than half in this century alone.

The plight of the Amazonian rain forest has raised concern among many South Americans as well as people throughout the world. One of the rain forest's greatest champions was a Brazilian named Chico Mendes who on numerous occasions confronted and drove off workers hired by cattle ranchers to clear areas of the forest. Through his activism, Mendes made some strong enemies, one of whom gunned him down outside his home in 1988.

There are those who argue that, in order to grow economically and support its expanding population, South America must make full use of the lands of the Amazon basin. They point out that many developed countries are guilty of similar environmental exploitation at home.

The irony of the destruction of the Amazonian rain forest is that once the land is cleared of its native plants and trees, it is ill-suited for the demands of crops. After just a few years, the soil's fertility is exhausted. The people who cleared the land soon abandon it, leaving a landscape that has been robbed of its biodiversity.

In recent years, Brazil and other countries have strengthened legislation aimed at protecting the rain forest. However, they lack the resources to effectively enforce the laws, and today destruction of the rain forest continues at an alarming rate.

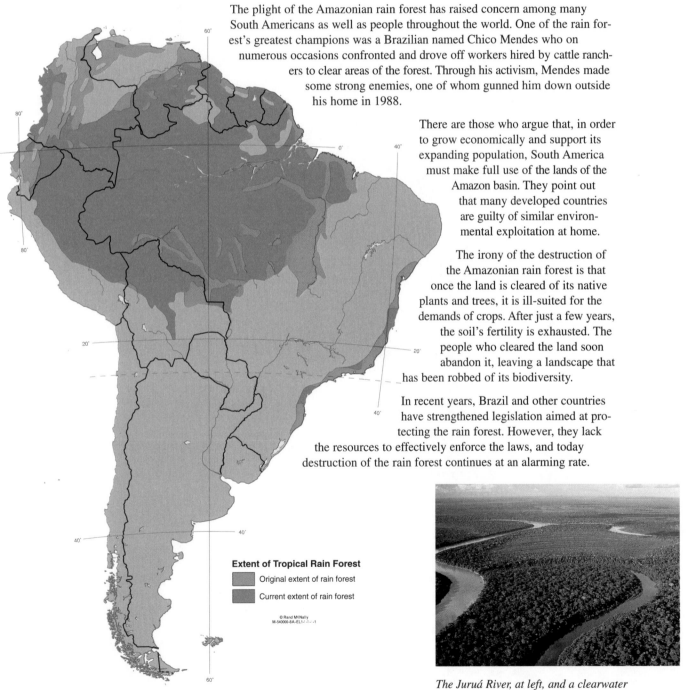

Extent of Tropical Rain Forest

Original extent of rain forest

Current extent of rain forest

© Rand McNally
M-540000-8A-EL1-.-.-.-1

The Juruá River, at left, and a clearwater slough wind through the dense Amazon rain forest near Eirunepé, Brazil.

HAVANA

Bahía de Campeche

PEN. DE YUCATÁN

CUBA

Yucatán Channel

JAMAICA

Gulf of Honduras

CENTRAL

AMERICA

Lago de Nicaragua

Panamá

ISTMO DE PANAMÁ

Golfo de Panamá

ISLA DEL COCO (Costa Rica)

ISLA DE MALPELO (Colombia)

ARCHIPIÉLAGO DE COLÓN (GALÁPAGOS ISLANDS) (Ec.)

HISPANIOLA

CARIBBEAN SEA

PUNTA DE GALLINAS

Golfo de Venezuela

San Juan

PUERTO RICO TRENCH

PUERTO RICO (U.S.A.)

GUADELOUPE (Fr.)

MARTINIQUE (Fr.)

BARBADOS

TRINIDAD AND TOBAGO

Port of Spain

NORTH AMERICAN BASIN

Tropic of Cancer

ATLANTIC OCEAN

WEST INDIES

Windward Passage

Barranquilla

Cartagena

Maracaibo

La Guaira

Valencia

CARACAS

Mérida

Ciudad Bolívar

Golfo del Darién

Medellín

BOGOTÁ

COLOMBIA

Nevado del Tolima 17,110

VENEZUELA

LLANOS

Cerro Icutú 7600

Orinoco

GUYANA

Georgetown

Paramaribo

SURINAME

Cayenne

FR. GUIANA

Boa Vista do Rio Branco

GUIANA HIGHLANDS

Quito

ECUADOR

Cotopaxi 19,347

Chimborazo 20,702

Guayaquil

Golfo de Guayaquil

Iquitos

Chiclayo

Trujillo

Nevs. Huascarán 22,133

LIMA

Callao

Cusco

Arequipa

Mollendo

Volcán Misti

PERU

ANDES MTS.

Letícia

Putumayo

Napo

Marañón

Ucayali

Juruá

Purús

Guaviare

Negro

Japurá

Amazon (Amazonas) (Solimões)

Madeira

Manaus (Manáos)

Porto Velho

Río Branco

Roosevelt

Tapajós

Xingú

Tocantins

ILHA DE MARAJÓ

Belém (Pará)

São Luís (Maranhão)

Equator

ROCEDOS SÃO PEDRO E SÃO PAULO (Brazil)

Fortaleza (Ceará)

ARQUIPÉLAGO FERNANDO DE NORONHA (Brazil)

CABO DE SÃO ROQUE

Teresina

Natal

João Pessoa (Paraíba)

RECIFE (Pernambuco)

Maceió

BRAZIL

CHAPADA DE MATO GROSSO

Cuiabá

SERRA DO PIAUÍ

BRAZILIAN HIGHLANDS

Salto

Rio São Francisco

Salvador (Bahia)

La Paz

Nev. Illimani 20,741

BOLIVIA

Sucre

Potosí

Lago de

Iquique

Antofagasta

ISLA DE SAN FÉLIX ISLA DE SAN AMBROSIO (Chile)

Cerro Azul 19,947

Copiapó

Coquimbo

Valparaíso

SANTIAGO

ISLAS DE JUAN FERNÁNDEZ (Chile)

Concepción

Valdivia

Puerto Montt

ISLA DE CHILOE

ARCHIPIÉLAGO DE LOS CHONOS

WELLINGTON

HANOVER

Punta Arenas

DESOLACIÓN

Mt. Sarmiento 8100

Tropic of Capricorn

Aconcagua 22,831

DESIERTO DE ATACAMA

PERU-CHILE TRENCH

C H I L E

Salta

Tucumán

Corrientes

Córdoba

Mendoza

Rosario

Santa Fe

La Plata

BUENOS AIRES

PAMPAS

Bahía Blanca

Viedma

Golfo San Matías

Río Colorado

Río Negro

Monte Valentín 5314

Comodoro Rivadavia

Golfo San Jorge

A R G E N T I N A

GRAN CHACO

Pilcomayo

Bermejo

Paraguay

PARAGUAY

Asunción

Paraná

Iguassú Falls

URUGUAY

Salto

MONTEVIDEO

Rio de la Plata

Rio Grande

Porto Alegre

Florianópolis

Santos

CABO FRIO

RIO DE JANEIRO

Vitória

Pico da Bandeira 9482

SERRA DO ESPINHAÇO

Belo Horizonte

Diamantina

Brasília

Paraná

SÃO PAULO

Chaco

Diablo

ATLANTIC OCEAN

PACIFIC OCEAN

FALKLAND IS. (ISLAS MALVINAS) (Br.)

Stanley

Río Gallegos

Estrecho de Magallanes

TIERRA DEL FUEGO

ISLA DE LOS ESTADOS

CABO DE HORNOS (CAPE HORN)

Drake Passage

SOUTH GEORGIA (Br.)

SOUTH ORKNEY IS. (Br.)

SOUTH SANDWICH ISLANDS (Br.)

SOUTH SANDWICH TRENCH

SOUTH SHETLAND ISLANDS (Br.)

JOINVILLE

ANTARCTIC PENINSULA

JAMES ROSS

Antarctic Circle

A-540000-26-4-4-7-16

COPYRIGHT BY RAND McNALLY & COMPANY MADE IN U.S.A.

Longitude West of Greenwich

40,000 SQ MI AREA

0   300   600

Miles

0   200   400   600   800   1000   Miles

0   400   800   1200   1600   Kilometers

Scale 1:40 000 000; one inch to 630 miles. Lambert's Azimuthal, Equal Area Projection
Elevations and depressions are given in feet

Tropic of Cancer

HAVANA
PEN. DE YUCATÁN
Golfo de Campeche
Lago de Nicaragua
ISLA DE MALPELO (Colombia)
ISLA DEL COCO (Costa Rica)
HISPANIOLA
San Juan
PUERTO RICO (U.S.A.)
JAMAICA
WEST INDIES
CARIBBEAN SEA
PUNTA DE GALLINAS
GUADELOUPE (Fr.)
MARTINIQUE (Fr.)
BARBADOS
TRINIDAD AND TOBAGO
Port of Spain

NORTH AMERICAN BASIN

ATLANTIC OCEAN

CENTRAL AMERICA
Panamá
Golfo de Panamá

Barranquilla
Cartagena
Maracaibo
Valencia
La Guaira
CARACAS
Ciudad Bolívar
VENEZUELA
Georgetown
Paramaribo
GUYANA
Cayenne
SURINAME
FR. GUIANA
Boa Vista do Rio Branco
GUIANA HIGHLANDS

Medellín
BOGOTÁ
COLOMBIA

Quito
ECUADOR
Cotopaxi
Guayaquil
Iquitos
Leticia
Manaus (Manáos)

ARCHIPIÉLAGO DE COLÓN (GALÁPAGOS ISLANDS) (Ec.)

ILHA DE MARAJÓ
Belém (Pará)
São Luís (Maranhão)
Equator
ROCEDOS SÃO PEDRO E SÃO PAULO (Brazil)

Chiclayo
Trujillo
Nevs. Huascarán 22,133
PERU
ANDES
Cusco
Callao
LIMA
Volcán Misti
Arequipa
Mollendo

Amazon (Amazonas)
Fortaleza (Ceará)
Teresina
Natal
João Pessoa (Paraíba)
RECIFE (Pernambuco)
Maceió
CABO DE SÃO ROQUE
ARQUIPÉLAGO FERNANDO DE NORONHA (Brazil)

B R A Z I L
CHAPADA DE MATO GROSSO
Cuiabá
BRAZILIAN HIGHLANDS
Salvador (Bahia)

La Paz
BOLIVIA
Sucre
Potosí
Brasília
Belo Horizonte
Diamantina
Pico da Bandeira
Vitória

Tropic of Capricorn
Antofagasta
ISLA DE SAN FÉLIX (Chile)
ISLA DE SAN AMBROSIO (Chile)
Copiapó
GRAN CHACO
PARAGUAY
Asunción
SÃO PAULO
Santos
CABO FRIO
RIO DE JANEIRO

Coquimbo
ARGENTINA
Tucumán
Corrientes
Florianópolis

Valparaíso
SANTIAGO
CHILE
Córdoba
Santa Fe
Salto
Rosario
URUGUAY
Río Grande
Porto Alegre

ISLAS DE JUAN FERNÁNDEZ (Chile)
Mendoza
BUENOS AIRES
La Plata
MONTEVIDEO

Concepción
PAMPAS
Río de la Plata
Bahía Blanca

Valdivia
Viedma
Golfo San Matías

Puerto Montt
ISLA DE CHILOÉ
ARCHIPIÉLAGO DE LOS CHONOS

Comodoro Rivadavia
Golfo San Jorge
Monte Valentín 13,314

WELLINGTON
HANOVER
Punta Arenas
DESOLACIÓN
Mt. Sarmiento 8100
TIERRA DEL FUEGO
ISLA DE LOS ESTADOS
CABO DE HORNOS (CAPE HORN)

FALKLAND IS. (ISLAS MALVINAS) (Br.)
Río Gallegos
Stanley
Estrecho de Magallanes

SOUTH GEORGIA (Br.)

ATLANTIC OCEAN

PACIFIC OCEAN

Drake Passage

SOUTH SHETLAND ISLANDS (Br.)
SOUTH ORKNEY IS. (Br.)
JOINVILLE
JAMES ROSS
ANTARCTIC PENINSULA
SOUTH SANDWICH ISLANDS (Br.)
Antarctic Circle

A-540000-76
COPYRIGHT BY
RAND McNALLY & COMPANY
MADE IN U.S.A.

Longitude West of Greenwich

Relief		
Meters		Feet
3050		10 000
1525		5000
610		2000
305		1000
0	Sea Level	0
152.5		500
1525		5000
3050		10 000
6100		20 000

| 0 | 200 | 400 | 600 | 800 | 1000 Miles |
| 0 | 400 | 800 | 1200 | 1600 Kilometers |

Scale 1:40 000 000; one inch to 630 miles. Lambert's Azimuthal, Equal Area Projection
Elevations and depressions are given in feet

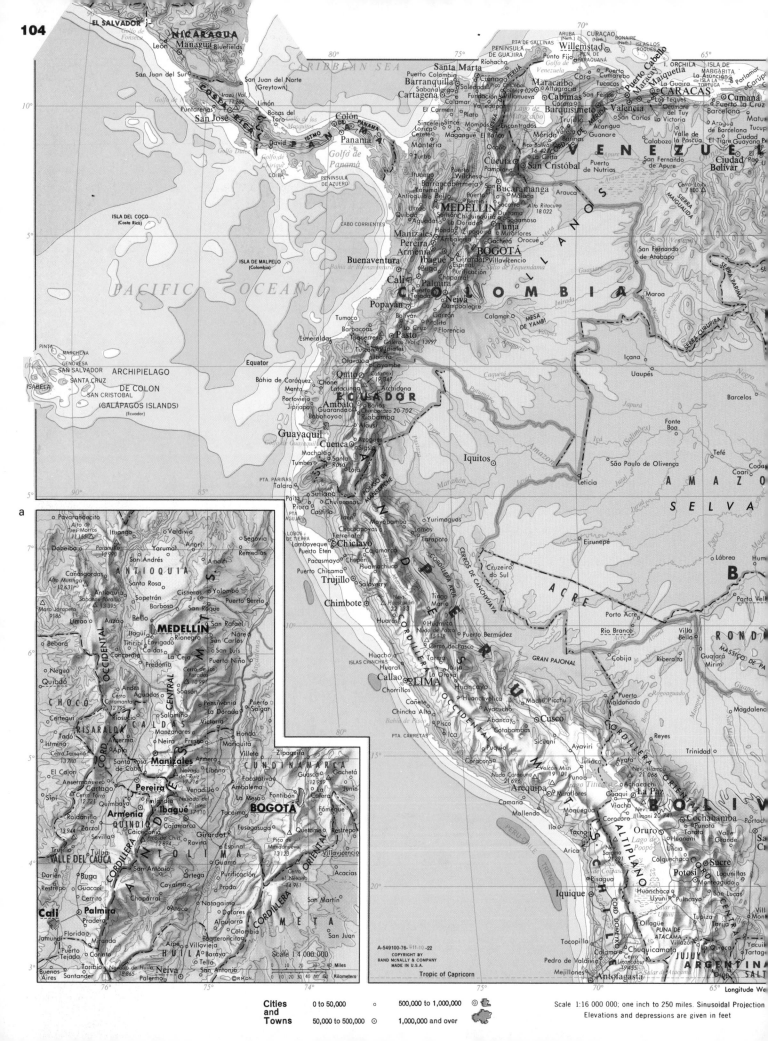

Relief

Meters		Feet
3050		10 000
1525		5000
610		2000
305		1000
152.5		500
0	Sea Level	0
152.5		500
1525	Below Sea Level	5000
3050		10 000
6100		20 000

0   50 100   200   300   400   500 Miles

0   100   200   300   400   500   600   700   800 Kilometers

Scale 1:16 000 000; one inch to 250 miles. Sinusoidal Projection
Elevations and depressions are given in feet

a

BUENOS AIRES

Scale 1:1 000 000

0   2   4   6   8   10 Miles

0 2 4 8   12   16 Kilometers

©RMCN.

b

RIO DE JANEIRO

Scale 1:1 000 000

0   2   5   10 Miles

0 2 4 8   12   16 Kilometers

©RMCN.

A-549200-76 -11- -14
COPYRIGHT BY
RAND McNALLY & COMPANY
MADE IN U.S.A.

Longitude West of Greenwich

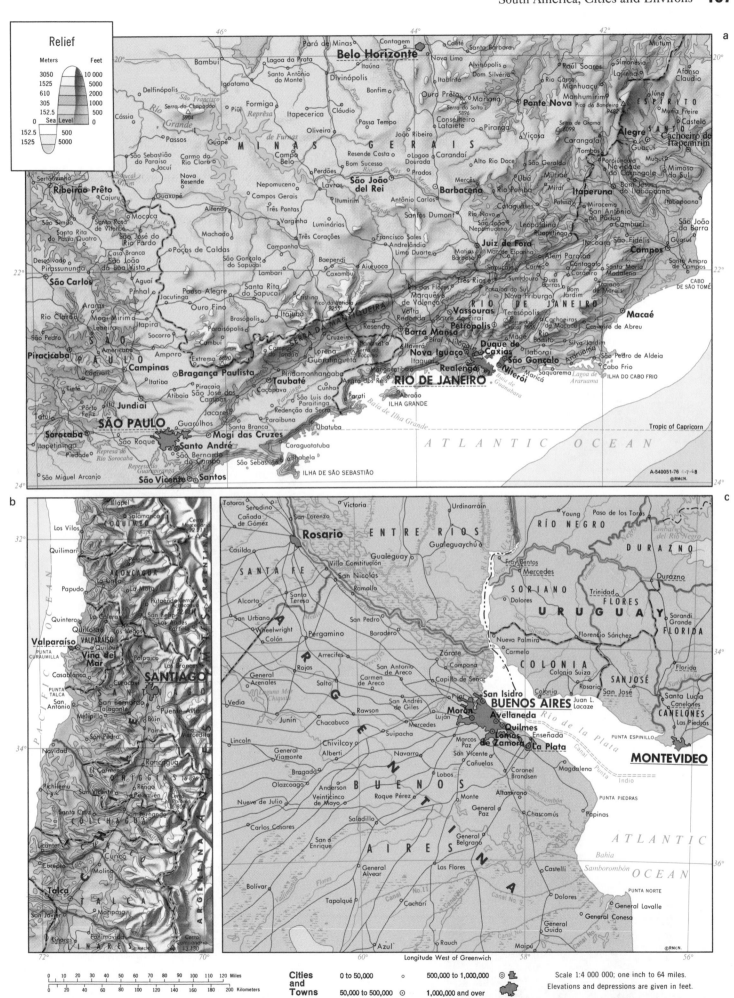

Europe

Europe is smaller than every other continent except Australia. In a sense, Europe is not really a continent at all, since it is part of the same vast landmass as Asia. Geographers sometimes refer to this landmass as a single continent, Eurasia. Europe occupies only about 18% of the land area of Eurasia.

Europe can be described as an enormous peninsula, stretching from the Ural Mountains, Ural River, and Caspian Sea in the east, to the Atlantic Ocean in the west; and from the Arctic Ocean in the north to the Mediterranean Sea, Black Sea, and Caucasus mountains in the south. The British Isles, Iceland, Corsica, Crete, and thousands of smaller islands that lie off the European mainland are usually considered as part of the continent.

A sweep of mountain ranges, including the Pyrenees, Alps and Carpathians, divides the colder, wetter north from the sun-drenched south.

Europe at a glance

Land area: 3,800,000 square miles (9,900,000 sq km)

Estimated population: 728,975,000

Population density: 192/square mile (74/sq km)

Mean elevation: 980 feet (300 m)

Highest point: Gora El' brus, Russia, 18,510 feet (5,642 m)

Lowest point: Caspian Sea, Asia-Europe, 92 feet (28 m) below sea level

Longest river: Volga, 2,194 mi (3,531 km)

Number of countries (incl. dependencies): 49

Largest independent country: Russia (Europe/Asia), 6,592,849 square miles (17,075,400 sq km)

Smallest independent country: Vatican City, 0.2 square miles (0.4 sq km)

Most populous independent country: Russia (Europe/Asia), 145,215,000

Least populous independent country: Vatican City, 1,000

Largest city: Moscow, pop. 8,368,449

Landforms

- Mountains
- Widely spaced mountains
- High tablelands
- Hills and low tablelands
- Plains
- Depresssions, basins
- High tablelands and ice caps
- Mountains and ice caps

© Rand McNally & Co.
M-550000-7C-EL1-

The Alps tower above a village in the Virgen Tal valley of western Austria.

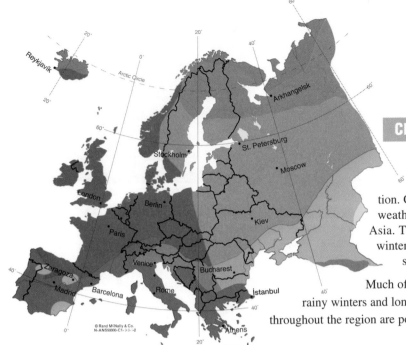

Climate

Warm, moist air masses flowing in from the Atlantic Ocean give much of Europe a mild climate and abundant precipitation. Cities like London, Paris and Rome all enjoy warmer weather than cities at similar latitudes in North America and Asia. The moderate winds don't reach eastern Europe, where the winters are long and cold and the summers short and cool. The same is true in the northern regions of Scandinavia.

Much of the south enjoys a Mediterranean climate, marked by short, rainy winters and long, dry summers. Indeed, the many beaches and islands found throughout the region are popular with vacationers year-round.

Tinted areas show temperature in degrees Fahrenheit. Vertical bars show precipitation in inches.

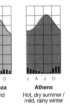 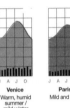 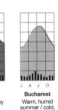 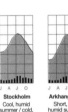 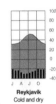

Zaragoza	Athens	Venice	Paris	Bucharest	Stockholm	Arkhangelsk	Reykjavik	Extensive uplands
Semiarid	Hot, dry summer / mild, rainy winter	Warm, humid summer / mild winter	Mild and rainy	Warm, humid summer / cold, snowy winter	Cool, humid summer / cold, snowy winter	Short, cool, humid summer / very cold, snowy winter	Cold and dry	Climate varies with elevation and latitude

Population

Europe is the second most densely populated continent. Only Asia has a greater population density. However, Europe's density varies dramatically from country to country. The Netherlands, for instance, has a density of 991 people per square mile (383 per sq km), making it one of the most densely populated countries in the world. In contrast, Norway has only 30 people per square mile (12 per sq km).

A vast array of ethnic groups and cultures can be found in Europe's relatively small area. Throughout the centuries, this diversity has enriched European culture while also leading to many hostilities. Of the 60 languages spoken, the majority are derived from Latin, Germanic or Slavic roots. Most Europeans are Christian, either Protestant or Roman Catholic.

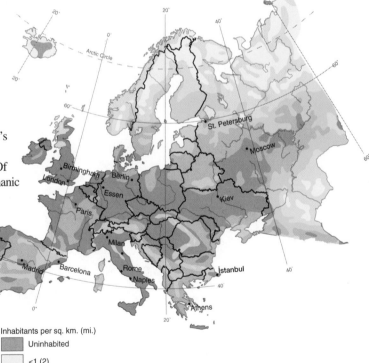

Fastest-Growing Countries

Croatia, 1.48% (annual rate of natural increase)

San Marino, 1.45%

Bosnia and Herzegovina, 1.38%

Slowest-Growing Countries

Bulgaria, -1.14%

Latvia, -0.81%

Ukraine, -0.78%

Inhabitants per sq. km. (mi.)

- Uninhabited
- <1 (2)
- 1-10 (2-25)
- 10-25 (25-60)
- 25-50 (60-125)
- 50-100 (125-250)
- >100 (250)

© Rand McNally & Co.
M-550000-1P-EL1-

Environments and Land Use

Given the high population density of Europe, it is not surprising that evidence of human development can be seen in every part of the continent, with the exception of the northern reaches of Scandinavia. In Western Europe, small farms are surrounded by towns, cities and industrial areas. Only in the east, in areas such as the vast rolling steppes of the Ukraine, can large farms and unbroken natural vistas be found.

Harvesting grapes from vineyards in Burgundy, France

The heavily industrialized countries of Western Europe boast rich economies and high standards of living. Switzerland has a per capita Gross Domestic Product (GDP) approaching U.S. $22,000, the highest in the world. The figures are generally much lower in Eastern Europe. One of the continent's poorest countries is the tiny former Communist state of Albania, which has a per capita GDP of only U.S. $998.

Pollution is an unfortunate by-product of the continent's industry. One example is the scenic Rhine River: in the 1980s, large stretches were found to be so polluted that they were devoid of life. These findings sparked a 20-year program to clean up the river. In general, the countries of Eastern Europe suffer from the worst pollution, as economic development has in the past taken precedence over environmental policies.

The vast forests of Scandinavia support a large paper and wood-products economy. Where forests survive in countries farther to the south, they are often used for recreation. Along the Mediterranean, the warm and dry lands support olive and fruit orchards. In many of these areas, agriculture is being supplemented and even replaced by tourism.

Over-fishing has depleted the seas and ocean around Europe. The fleets of countries such as Spain and Great Britain must sail far into the North Atlantic to find the ever-dwindling stocks of fish.

Urban
Cropland
Cropland and woodland
Cropland and grazing land
Grassland, grazing land
Forest, woodland
Swamp, marsh
Tundra
Shrub, sparse grass, wasteland
Barren land

© Rand McNally & Co.
M-550000-8L-EL1-1-1-1-1

Political Changes Since 1989

Much of Europe lay in economic and physical ruin after World War II ended in 1945. Germany's cities and industrial centers had been ravaged by aerial bombardment and assaults by the Allied armies. Many other countries, such as Russia, Poland, Belgium, and the Netherlands suffered gravely from Nazi invasions and occupation.

After 1945, tensions among the victorious Allies grew, specifically between the western powers—the United States, Great Britain and France—and the Soviet Union. It became clear that the two sides had vastly different visions for post-war Europe. In 1946, former British prime minister Winston Churchill observed that an "Iron Curtain" had gone down across Europe. It was to stay in place for almost 45 years.

Germany and the city of Berlin were divided between the western powers and the Soviet Union. West Germany quickly joined the other countries of Western Europe in building a stable, affluent democratic society. East Germany became part of a bloc of Eastern European countries dominated by the Soviet Union. These included Poland, Czechoslovakia, Hungary, Romania, Albania, and Bulgaria (see map at right). The economies in these countries were tightly controlled and personal freedoms were severely limited by the Communist governments in power.

The two blocs faced each other in a tense, generally non-military standoff called "The Cold War," which lasted for four decades. However, in 1985 winds of reform began sweeping through Eastern Europe. In 1989, Hungary relaxed its borders with Austria, setting off a flow of refugees from the east who had been forbidden to travel to the west. Thus began a dizzying period of change: the next two years saw the collapse of the Soviet Union, the reunification of Germany, independence for the former Soviet republics, and freedom from Soviet influence for the former bloc countries.

Although much of the old east seems intent on adopting western ideals of democracy and freedom, the process is not without problems. Switching from communism to market economies has meant hardship for millions. It has also led to ethnic tensions that resulted in the peaceful break-up of Czechoslovakia and the violent civil wars in the former Yugoslavia.

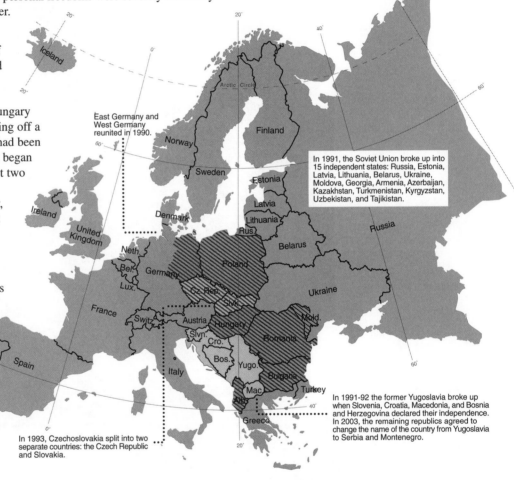

East Germany and West Germany reunited in 1990.

In 1991, the Soviet Union broke up into 15 independent states: Russia, Estonia, Latvia, Lithuania, Belarus, Ukraine, Moldova, Georgia, Armenia, Azerbaijan, Kazakhstan, Turkmenistan, Kyrgyzstan, Uzbekistan, and Tajikistan.

In 1993, Czechoslovakia split into two separate countries: the Czech Republic and Slovakia.

In 1991-92 the former Yugoslavia broke up when Slovenia, Croatia, Macedonia, and Bosnia and Herzegovina declared their independence. In 2003, the remaining republics agreed to change the name of the country from Yugoslavia to Serbia and Montenegro.

Political Change Since 1989

- Former Soviet Union
- Former Czechoslovakia
- Former Yugoslavia
- Former East and West Germany
- Former Soviet-bloc countries

© Rand McNally & Co.
M-550000-2P-EL1-·-·- -2

Berlin Wall memorial

40,000 SQ MI
AREA

0   100   200
Miles

Scale 1: 16 000 000; one inch to 250 miles. Conic Projection
Elevations and depressions are given in feet

Longitude West of Greenwich   Longitude East of Greenwich

0   50   100   200   300   400   500 Miles
0   100   200   400   600   800 Kilometers

Relief

Meters		Feet
3050		10 000
1525		5000
610		2000
305		1000
152.5		500
0	Sea Level	0
152.5		Below Sea Level
1525	500	
3050	5000	
	10 000	

Scale 1: 16 000 000; one inch to 250 miles. Conic Projection
Elevations and depressions are given in feet

Longitude West of Greenwich   Longitude East of Greenwich

| 0 | 50 | 100 | 200 | 300 | 400 | 500 Miles |

| 0 | 100 | 200 | 400 | 600 | 800 Kilometers |

Relief

Meters	Feet	
3050	10 000	
1525	5000	
610	2000	
305	1000	
152.5	500	
0	0	Sea Level
152.5	500	Below Sea Level
1525	5000	
3050	10000	

Scale 1: 10 000 000; one inch to 160 miles. Conic Projection
Elevations and depressions are given in feet

RUSSIA · FINLAND · ESTONIA · LATVIA · LITHUANIA · BELARUS

SWEDEN · NORWAY · DENMARK

Murmansk · Polyarnyy · Vardø · Vadsø · Hammerfest · Nordkapp · Alta · Kirkenes · Kola

Oulu · Kemi · Tornio · Rovaniemi · Kuusamo · Kajaani · Kuopio · Jyväskylä · Tampere · Turku · Helsinki · Lahti · Kotka · Pori · Rauma · Vaasa

Tallinn · Riga · Liepāja · Klaipėda · Kaunas · Šiauliai · Panevėžys · Jelgava · Kaliningrad · Gdańsk · Gdynia

STOCKHOLM · Uppsala · Gävle · Norrköping · Linköping · Örebro · Västerås · Eskilstuna · Jönköping · Göteborg · Helsingborg · Kalmar · Visby · Karlskrona · Kristianstad · Borås · Karlstad · Östersund · Sundsvall · Härnösand · Hudiksvall · Söderhamn · Kiruna · Gällivare · Luleå · Piteå · Umeå · Skellefteå · Boden

Trondheim · Oslo · Bergen · Stavanger · Haugesund · Egersund · Kristiansand · Arendal · Molde · Kristiansund · Narvik · Bodø · Mo i Rana · Nesna · Namsos · Lillehammer · Hamar · Drammen · Larvik · Skien · Lindesnes

COPENHAGEN (København) · Odense · Ålborg · Århus · Esbjerg · Randers · Kolding · Helsingør · Roskilde · Frederikshavn · Flensburg

UNITED KINGDOM · SCOTLAND · NORTHERN IRELAND · IRELAND · BRITISH ISLES

Aberdeen · Dundee · Edinburgh · GLASGOW · Greenock · Paisley · Perth · Inverness · Wick · Dornoch · Stornoway · Newcastle upon Tyne · South Shields · Sunderland · Hartlepool · Middlesbrough · Carlisle · Barrow-in-Furness · MANCHESTER · Belfast · Londonderry · Dundalk · Drogheda · Dublin (Baile Átha Cliath) · Sligo

SHETLAND IS. (Br.) · Lerwick · Mainland · ORKNEY IS. (Br.) · Kirkwall · HEBRIDES · ISLE OF SKYE · TIREE · ISLAY · MULL OF KINTYRE · ACHILL ISLAND · SLYNE HEAD · MALIN HD.

FAROE IS. (Dem.) · Tórshavn · JAN MAYEN (Nor.)

ICELAND · Reykjavik · Hafnarfjördur · Akureyri · Siglufjördur · Seydisfjördur · Eskifjördur · Vestmannaeyjar · Vopnafjördur

ARCTIC OCEAN · NORWEGIAN SEA · NORTH SEA · BALTIC SEA · GULF OF BOTHNIA · Gulf of Finland · Gulf of Riga · Kattegat · Skagerrak · DOGGER BANK

Arctic Circle

Longitude East of Greenwich
Longitude West of Greenwich
Miles
Kilometers
A-559400-76 -13-10-26
COPYRIGHT BY RAND McNALLY & COMPANY
MADE IN U.S.A.
MEDITERRANEAN SEA
TYRRHENIAN SEA
ADRIATIC SEA
IONIAN SEA
LIGURIAN SEA
BAY OF BISCAY
ENGLISH CHANNEL
FRANCE
SPAIN
GERMANY
ITALY
TUNISIA
ATLAS MOUNTAINS
PARIS
LONDON
MADRID
ROME
LISBON
BARCELONA
MILAN
NAPLES
BRUSSELS
AMSTERDAM
BERLIN
PRAGUE
VIENNA
BUDAPEST
Algiers
Tunis
CORSICA (Fr.)
SARDINIA (It.)
ILLES BALEARS (Sp.)

Scale 1:10 000 000; one inch to 160 miles. Conic Projection

Elevations and depressions are given in feet.

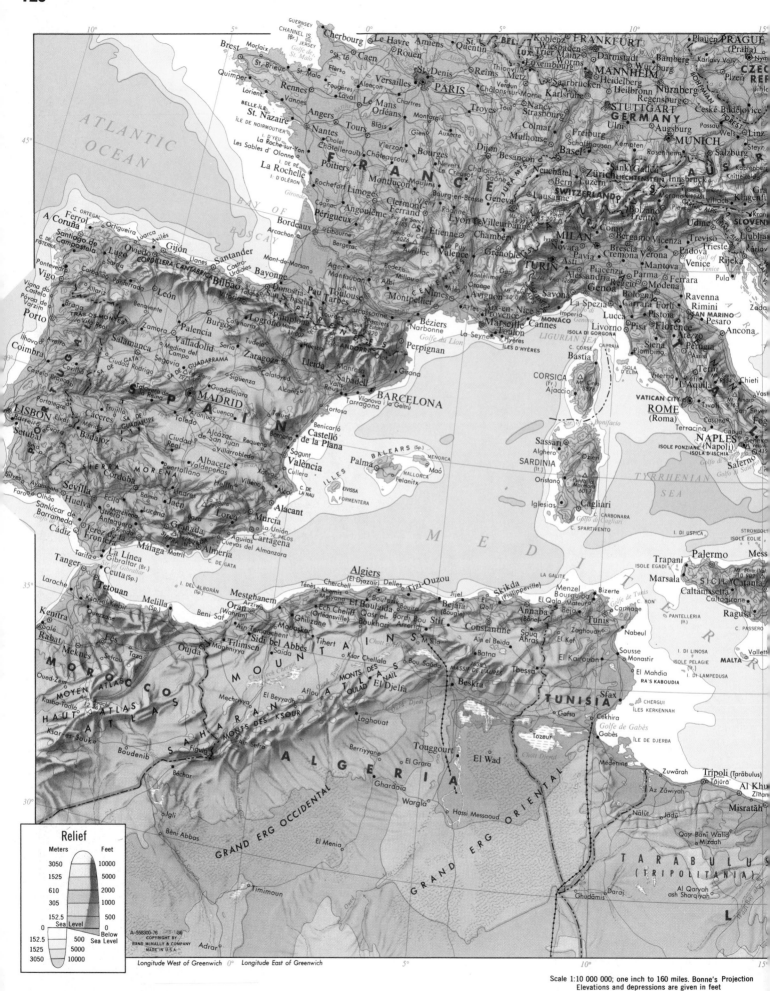

Relief

Meters		Feet
3050		10000
1525		5000
610		2000
305		1000
152.5		500
Sea Level		Sea Level
152.5		500 Below
1525		5000 Sea Level
3050		10000

A-558300-76
COPYRIGHT BY
RAND McNALLY & COMPANY
MADE IN U.S.A.

Longitude West of Greenwich 0° Longitude East of Greenwich

Scale 1:10 000 000; one inch to 160 miles. Bonne's Projection
Elevations and depressions are given in feet

Africa

The Drakensberg Mountains in southern South Africa.

Africa, the second-largest continent, comprises about one-fifth of the world's land area. From the Equator, Africa extends roughly the same distance to the north as it does to the south.

A high plateau covers much of the continent. The edges of the plateau are marked by steep slopes, called escarpments, where the land angles sharply downward onto narrow coastal plains or into the sea. Many of the continent's great rivers plunge over these escarpments in falls or rapids, and therefore cannot be used as transportation routes from the coast into the continent's interior.

Among Africa's most significant mountain systems are the Atlas range in the far north and the Drakensberg range in the far south. A long string of mountain ranges and highlands running north-south through eastern Africa marks the course of the Great Rift Valley.

Africa at a glance

Land area: 11,700,000 square miles (30,300,000 sq km)

Estimated population: 832,590,000

Population density: 71/ square mile (27/sq km)

Mean elevation: 1,900 feet (580 m)

Highest point: Kilimanjaro, Tanzania, 19,340 feet (5,895 m)

Lowest point: Lac Assal, Djibouti, 515 feet (157 m) below sea level

Longest river: Nile, 4,145 mi (6,671 km)

Number of countries (incl. dependencies): 61

Largest independent country: Sudan, 967,500 square miles (2,505,813 sq km)

Smallest independent country: Seychelles, 176 square miles (455 sq km)

Most populous independent country: Nigeria, 128,285,000

Least populous independent country: Seychelles, 80,000

Largest city: Cairo, pop. 6,801,000

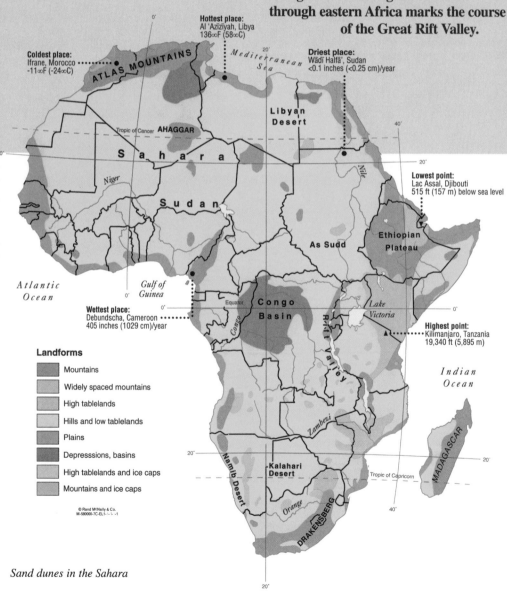

Coldest place: Ifrane, Morocco -11∞F (-24∞C)

Hottest place: Al 'Azīzīyah, Libya 136∞F (58∞C)

Driest place: Wādī Ḥalfā', Sudan <0.1 inches (<0.25 cm)/year

Lowest point: Lac Assal, Djibouti 515 ft (157 m) below sea level

Wettest place: Debundscha, Cameroon 405 inches (1029 cm)/year

Highest point: Kilimanjaro, Tanzania 19,340 ft (5,895 m)

ATLAS MOUNTAINS

Mediterranean Sea

Libyan Desert

Tropic of Cancer AHAGGAR

S a h a r a

Niger

S u d a n

Nile

As Sudd

Ethiopian Plateau

Atlantic Ocean

Gulf of Guinea

Equator

Congo Basin

Congo

Lake Victoria

Rift Valley

Indian Ocean

Zambezi

Namib Desert

Kalahari Desert

Orange

Tropic of Capricorn

DRAKENSBERG

MADAGASCAR

Landforms

- Mountains
- Widely spaced mountains
- High tablelands
- Hills and low tablelands
- Plains
- Depresssions, basins
- High tablelands and ice caps
- Mountains and ice caps

© Rand McNally & Co.
M-580000-7C-EL1-⅃ ⅃ ⅃ -1

Sand dunes in the Sahara

Climate

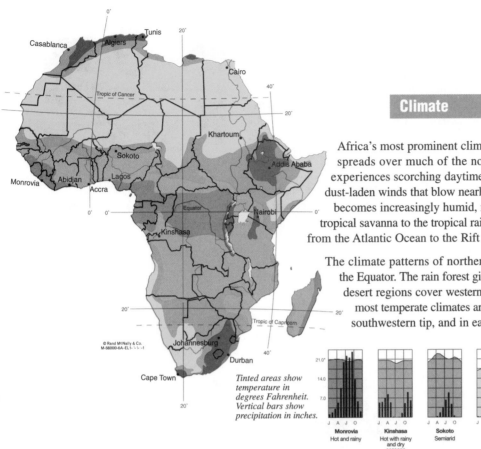

Africa's most prominent climatic region is the vast Sahara desert which spreads over much of the northern half of the continent. The Sahara experiences scorching daytime temperatures, minimal rainfall, and hot, dry, dust-laden winds that blow nearly continuously. South of the Sahara, the climate becomes increasingly humid, moving through zones of semiarid steppe and tropical savanna to the tropical rain forest that stretches across equatorial Africa from the Atlantic Ocean to the Rift Valley.

The climate patterns of northern Africa are repeated in reverse south of the Equator. The rain forest gives way to zones of decreasing humidity, and desert regions cover western South Africa and Namibia. Africa's mildest, most temperate climates are found along its Mediterranean coast, at its southwestern tip, and in eastern South Africa.

Tinted areas show temperature in degrees Fahrenheit. Vertical bars show precipitation in inches.

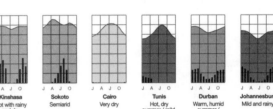

Monrovia	Kinshasa	Sokoto	Cairo	Tunis	Durban	Johannesburg	Extensive uplands
Hot and rainy	Hot with rainy and dry seasons	Semiarid	Very dry	Hot, dry summer / mild, rainy winter	Warm, humid summer / mild winter	Mild and rainy	Climate varies with elevation and latitude

© Rand McNally & Co.
M-580000-6A-EL1- -·- -1

Population

About one-seventh of the world's people live in Africa. It is the second most populous continent. The population is almost evenly divided between the sub-Saharan countries and those bordering the Mediterranean. Large tracts of the Sahara are uninhabited. Despite recurring famines, disease, and warfare, the population is rapidly increasing.

The largest concentrations of people are generally found in regions in which one or more of the following conditions exist: moderate temperatures, ample water supply, and arable land. These regions include Egypt's fertile Nile Valley, the northern coast of the Gulf of Guinea, the highlands of East Africa, and the coastal regions of Morocco, Algeria, and Tunisia, north of the Atlas Mountains.

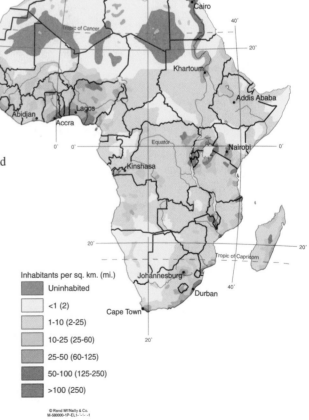

Inhabitants per sq. km. (mi.)
- Uninhabited
- <1 (2)
- 1-10 (2-25)
- 10-25 (25-60)
- 25-50 (60-125)
- 50-100 (125-250)
- >100 (250)

© Rand McNally & Co.
M-580000-1P-EL1- -·- -1

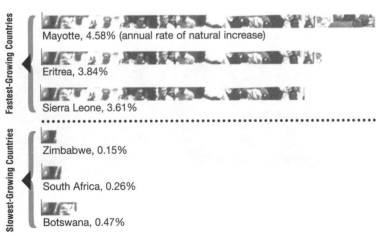

Fastest-Growing Countries

Mayotte, 4.58% (annual rate of natural increase)

Eritrea, 3.84%

Sierra Leone, 3.61%

Slowest-Growing Countries

Zimbabwe, 0.15%

South Africa, 0.26%

Botswana, 0.47%

Environments and Land Use

Deserts account for one-third of Africa's land area, and they claim new land every year. Drought, over-farming, and over-grazing can quickly turn marginal land, such as that of the Sahel region (also known as the Sudan), into barren waste-land. The huge Sahara desert has itself only existed a short time, in geological terms: cave paintings and other archeologi-cal evidence indicate that green pastureland covered the area just a few thousand years ago.

Shepherd with goats in the Sahel (Sudan)

Most Africans are subsistence farmers, growing sorghum, corn, millet, sweet potatoes, and other starchy foods. Commercial farms, most of which date from the colonial period, can be found throughout central and southern Africa, producing cash crops such as coffee, bananas, tobacco and cacao. One-quarter of the continent's land is suitable for grazing, but disease and drought have made raising animals difficult. Although three out of four Africans work in agriculture, Africa is the only continent that is not self-sufficient for food.

The great rain forests that cover much of equatorial Africa produce mahogany, ebony, and other valuable hardwoods. However, only limited areas of the forests are suit-able for logging, and the lack of developed road networks makes it difficult and costly to transport the wood.

Vast mineral reserves are spread throughout the continent. Most are unexploited, but notable excep-tions include the diamond mines of South Africa and Namibia, the cop-per mines of Zambia and Democratic Republic of the Congo, and the oil fields of Nigeria, Libya, and Algeria.

The great concentrations of wildlife for which Africa is famous can still be found in places such as Tanzania's Serengeti Plain and Botswana's Kalahari Desert. In many other parts of the conti-nent, however, wildlife is quickly disappearing as humans encroach on habitat and poachers decimate entire species.

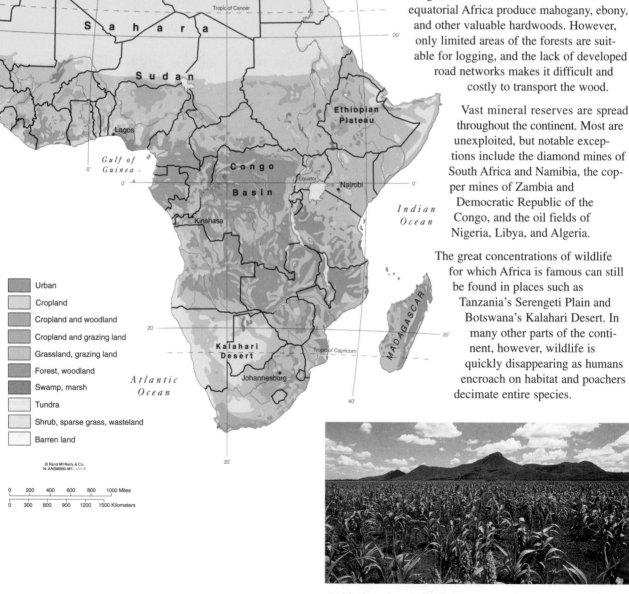

Urban

Cropland

Cropland and woodland

Cropland and grazing land

Grassland, grazing land

Forest, woodland

Swamp, marsh

Tundra

Shrub, sparse grass, wasteland

Barren land

© Rand McNally & Co.
N-ANS80000-M1- -2-2-1

0   200   400   600   800   1000 Miles

0   300   600   900   1200   1500 Kilometers

Field of sorghum in Zimbabwe

Africa: from Colonial Rule to Independence

The origins of Europe's colonization of Africa can be traced back to the 1500s, when a lucrative slave trade developed to supply European settlers in the New World with laborers. Africa became the primary source for slaves: between the mid-1500s and the mid-1800s, 11 million Africans were captured and sold into slavery.

When the slave trade was banned across Europe in the early 1800s, commercial trade with Africa continued. In the second half of the century, competition for Africa's minerals and other raw materials intensified, and between 1880 and 1914, France, Britain, Italy, Portugal, Belgium, Spain, and Germany annexed large areas of Africa. Colonial rule was often characterized by racial prejudice and segregation.

In the late 19th and early 20th centuries, Egypt, Ethiopia, and South Africa began to break free from colonial influence. For most or Africa, however, colonial rule persisted through the mid-1900s, although it faced growing bitterness and nationalist sentiment. As the colonial powers struggled through two world wars, and as their international dominance declined, it became increasingly difficult for them to maintain their empires.

In 1951, Libya gained its independence, following a UN resolution that ended British and French control. Sudan peacefully won independence from Britain and Egypt in 1956. A year later, Britain granted independence to the Gold Coast, which became the new country of Ghana. Guinea separated from France in 1958, followed by all of the other French colonies in 1960. Anti-colonial movements gathered strength across Africa, and by the end of the 1970s, a total of 43 countries had become independent.

The end of colonial rule, however, has not brought peace and prosperity. Many of the newly freed countries were ill-prepared for independence. Their economies were oriented to fit the needs of the now-departed colonists, few transportation networks existed, and dictators and rival despots fought for power in civil wars.

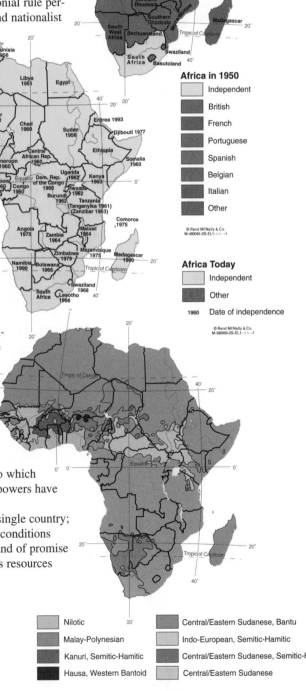

Africa in 1950
- Independent
- British
- French
- Portuguese
- Spanish
- Belgian
- Italian
- Other

© Rand McNally & Co.
M-480045-2S-EL1-⌐-⌐- -1

Africa Today
- Independent
- Other

1960   Date of independence

© Rand McNally & Co.
M-580000-2S-EL1-⌐-⌐- -1

People of the Samburu tribe

Further, most Africans identify themselves primarily with the tribe to which they belong. The political delineations established by the European powers have little meaning and often conflict with traditional tribal boundaries. In some cases, enemy tribes found themselves pushed together in a single country; in others, single tribes were divided among several countries. These conditions have already brought much warfare and hardship. Still, Africa is a land of promise and opportunity. The rich diversity of its people and abundance of its resources should inevitably enable the continent to realize its potential.

Ethno-linguistic Groups

Semitic-Hamitic	Bantu	Indo-European	Nilotic	Central/Eastern Sudanese, Bantu
Mande	Central Bantoid	Kanuri	Malay-Polynesian	Indo-European, Semitic-Hamitic
Guinean	Eastern Bantoid	Songhai	Kanuri, Semitic-Hamitic	Central/Eastern Sudanese, Semitic-Hamitic
Hausa	Western Bantoid	Khoisan	Hausa, Western Bantoid	Central/Eastern Sudanese

© Rand McNally & Co.
M-580000-1D-EL1-⌐-⌐- -1

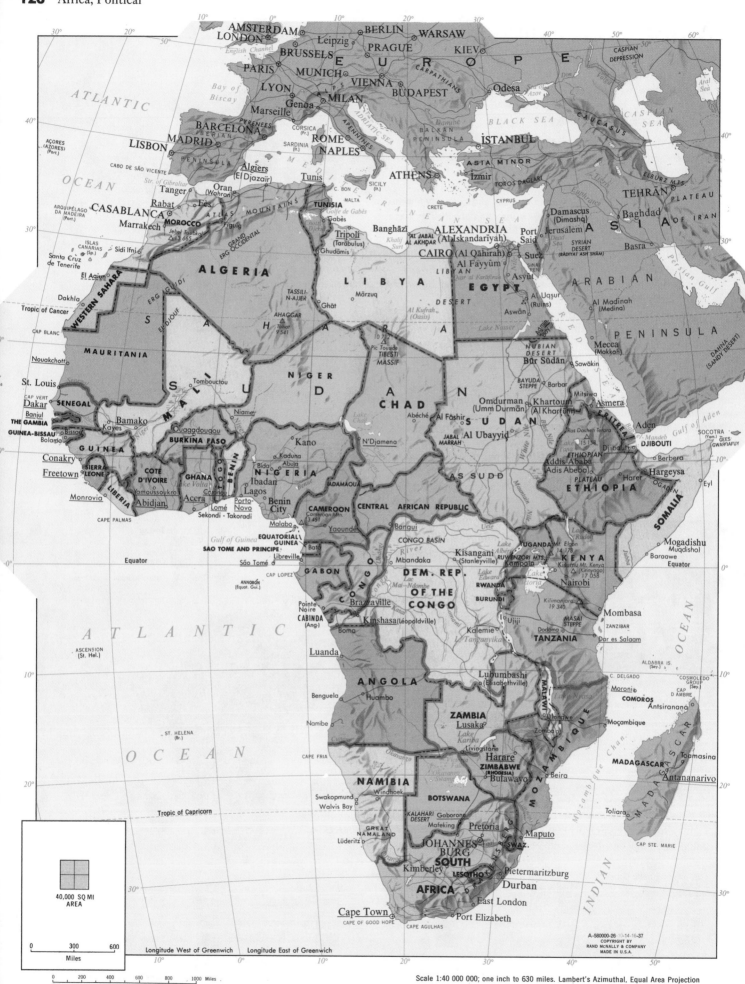

Scale 1:40 000 000; one inch to 630 miles. Lambert's Azimuthal, Equal Area Projection
Elevations and depressions are given in feet.

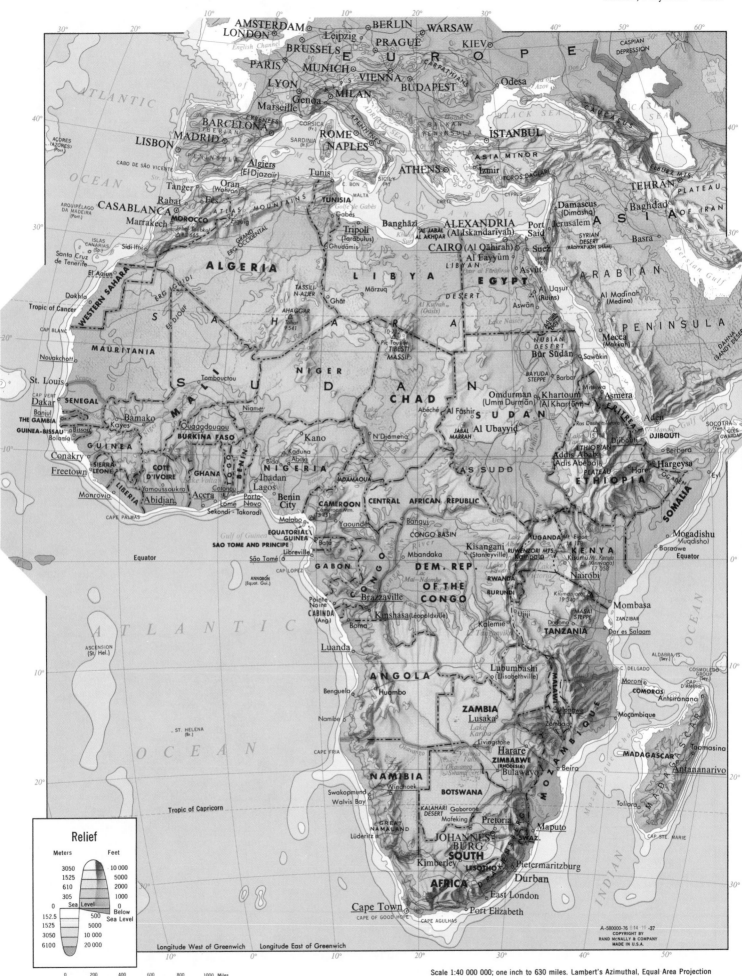

Relief

Meters		Feet
3050		10 000
1525		5000
610		2000
305		1000
0	Sea Level	0
		Below
		Sea Level
152.5		500
1525		5000
3050		10 000
6100		20 000

Longitude West of Greenwich   Longitude East of Greenwich

0   200   400   600   800   1000 Miles
0   400   800   1200   1600 Kilometers

Scale 1:40 000 000; one inch to 630 miles. Lambert's Azimuthal, Equal Area Projection
Elevations and depressions are given in feet.

A-580000-76 ⋅14 16 -37
COPYRIGHT BY
RAND McNALLY & COMPANY
MADE IN U.S.A.

a

®RMCN.
GRACIOSA
FAIAL · PICO · TERCEIRA
SÃO JORGE
AÇORES (AZORES) · SÃO MIGUEL
(Port.)
Ponta Delgada · STA. MARIA

Same scale as main map

ATLANTIC OCEAN

ARQUIPÉLAGO
ILHA DE PORTO SANTO
Funchal · ILHA DA MADEIRA
DA MADEIRA (Port.)

ISLAS CANARIAS (Sp.)
LA PALMA · TENERIFE · LANZAROTE
San Sebastián · Sta. Cruz de Tenerife · FUERTEVENTURA
GOMERA · GRAN CANARIA · CAP DRAA · C. YUBY
HIERRO · Las Palmas de Gran Canaria

SPAIN
Cádiz
Gibraltar (U.K.) · Algiers (El Djazair) · Delfes · Bejaïa (Bougie) · El Skikda · Annaba (Bône) · Tunis · Bizerte
Str. of Gibraltar · Ceuta (Sp.) · Mestghanem · Ech Cheliff · Cherchell · Tizi-Ouzou · Guelma · Ain el Beida
Tanger (Tangier) · Melilla (Sp.) · Oran · Ghilizane · Lemdiyya · El Boulaida · Stif · Batna · Souk · Kairot
Larache · Tetouan · Beni Saf · Sidi bel Abbès · Saïda · M'Sila · Ain el Beida · Tbessa · Sfa
Ouezzane · Oujda · Tilimsen · Beskra
Rabat · Fès · Taza
CASABLANCA · Meknès · Azemmour · Oued-Zem · Ain-Sefra · El Djelfa · Laghouat · El Wad · Touggourt · Gabès
El Jadida · Settat · Kasba-Tadla · Boudenib · ATLAS MOUNTAINS · Ghardaïa · Wargla · Ghudámi · AL H
Safi (Asfi) · MOROCCO · Demnat · Figuig · Hassi Messaoud · AL
Essaouira · Marrakech · Jebel-Toubkal 13665 · Béchar · GRAND ERG OCCIDENTAL · GRAND ERG ORIENTAL
Agadir · Taroudant · Igli · Timimoun · ALGERIA · Bordj Omar Idriss · In Amnas
ANTI ATLAS · Béni Abbès · El Menía · PLATEAU DU TADEMAÏT · In Salah · PLATEAU DU TINGHERT
Sidi Ifni · Tiznit · Tindouf · Adrar · Illizi
ERG IGUIDI · Chenachane · ERG CHECH · TIDIKELT · TASSILI-N-AJJER
El Aaiún · CABO BOJADOR · Ouallene · Ghât
WESTERN SAHARA · SAHARA · EL HANK · TANEZROUFT · Djanet
Dakhla · Tropic of Cancer · EL DJOUF · Taoudenni · Tahar 9541 · HAGGAR · Tamenghest

The Western Sahara is occupied by Morocco.
Fdérik · Taoudenni · Tamenghest

Nouâdhibou · CAP BLANC · CAP D'ARGUIN
Atar · Chinguetti · OUARANE · EL MREYYÉ · TUAREG · Mt. Gréboun 4562 · Ifrouâne
Nouâmrhar · CAP TIMIRIS · ADRAR DES IFÔGHAS · Monts Tamgak · AÏR
MAURITANIA · VALLÉE DU TILEMSI · Monts Tamak 3906
Nouakchott · Tidjikdja · Kidal · Monts Bagzane 6300
Boutilimit · El Mreyyé · Araouane · Agadez
Saint-Louis · Aleg · Kiffa · Néma · Oualâta · Tombouctou (Timbuktu) · Bamba · MALI · NIGER
Dagana · Kaédi · MBout · Néma · Goundam · Bourem · Gao
Podor · Matam · Sélibaby · Nioro du Sahel · Niafunké · Tahoua
Louga · Linguère · Nara · Tillabéry · Madaoua · Tessaoua · Zinder · Gouré
CAP VERT · Rufisque · Bakel · Goumbou · Sokolo · Niamey · Sokoto · Maradi · Nguru
Dakar · Thiès · Diourbel · Kayes · Bafoulabé · Ségou · San · Dori · Say · Dosso · Kaura Namoda · Katsina · Gumel · Geid
SENEGAL · Kaolack · Tambacounda · Kita · Koulikoro · Dienné · Mopti · Niamey · Birnin Kebbi · Gusau · Hadejia · BO · PL
THE GAMBIA · Banjul · M. du Tamgué 5046 · Satadougou · Bamako · Koutiala · Dédougou · BURKINA FASO · Fada Ngourma · Illo · Zaria · Kano · Potiskum
Ziguinchor · GUINEA-BISSAU · FOUTA DJALLON · Labé · Siguiri · Bougouni · Sikasso · Ouagadougou · Molanville · Kandi · Kontagora · Kaduna · Bauchi
Bissau · Bolama · Buba · Koumbia · Timbo · Mamou · Kankan · Bobo-Dioulasso · Koudougou · Tenkodogo · Gaoua · Gambaga · Sansanné-Mango · Natitingou · Zungeru · Minna · Jos · Gombe
ARQUIPÉLAGO DOS BIJAGÓS · Boké · GUINEA · Kouroussa · Odienné · KONG · Bouna · Yendi · Sokode · Parakou · Jebba · Abuja · NIGERIA
Boffa · Kindia · Faranah · Korhogo · Kong · Bole · Tamale · Keffi · Baro · Ibi
Forécariah · Kabala · Kissidougou · Beyla · Dabakala · Bondoukou · Kintampo · Savalou · Iseyin · Ilorin · Bida · Lokoja · Makurdi
Conakry · Makeni · KONG · Séguéla · GHANA · TOGO · Oyo · Oshogbo · Ilesha · Idah · Katsina Ala
SIERRA LEONE · Pandembu · Kolahun · Mont Nimba 5760 · Bouaké · Kumasi · Abomey · Ibadan · Ife · Benin City · Enugu
Freetown · Moyamba · Bouaflé · Lake Volta · Ada · Abeokuta · Ijebu Ode · Sapele · Onitsha · Fumban
Bonthe · Bomi Hills · COTE D'IVOIRE · Yamoussoukro · Koforidua · Lagos · Porto-Novo · Warri · Oweri · Mamfe · Dschang
Robertsport · Monrovia · (IVORY COAST) · Abidjan · Port-Bouet · Accra · Aba · Port Harcourt · Kumba · CAMEROON
LIBERIA · Buchanan · Grand Lahou · Grand Bassam · Assini · Sekondi-Takoradi · Cameroon Mtn. 13451 · Limbe · Douala · Yaoundé
River Cess · Greenville · Grand Cess · CAPE PALMAS · Tabou · C. THREE POINTS · Bonny · Malabo · BIOKO · Eséka
Harper · Bight of Benin · EQUATORIAL GUINEA · Bata · RIO MUNI · Edéa

ATLANTIC OCEAN · GULF OF GUINEA
SAO TOME AND PRINCIPE · ILHA DO PRÍNCIPE · Kribi · Ebolowa
ILHA DE SÃO TOMÉ · São Tomé · Libreville · GA

b

SANTA ANTÃO
SÃO VICENTE · SAL
SÃO NICOLAU · BOA VISTA
CAPE VERDE
SÃO TIAGO · MAIO
FOGO · Praia

Same scale as main map

A-589100-76 · 818-18-37EL
COPYRIGHT BY
RAND McNALLY & COMPANY
MADE IN U.S.A.
®RMCN.

Longitude West of Greenwich · Longitude East of Greenwich

Scale 1:16 000 000; one inch to 250 miles. Sinusoidal Projection
Elevations and depressions are given in feet

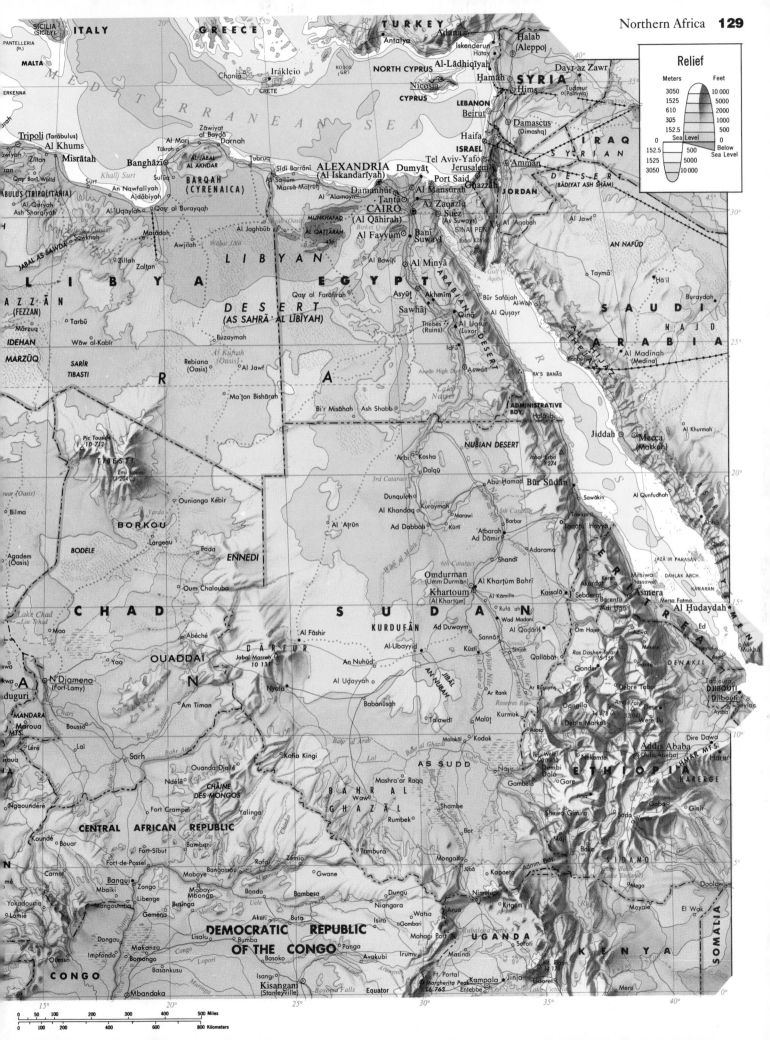

Relief

Meters	Feet
3050	10 000
1525	5000
610	2000
305	1000
152.5	500
0	Sea Level
152.5	500 Below Sea Level
1525	5000
3050	10 000

SICILIA (SICILY) • ITALY
PANTELLERIA (It.)
MALTA
ERKENNA

GREECE
Irákleio • Chaniá
CRETE
RODOS (GR.)
TURKEY
Antalya • Adana
Iskenderun • Hatay
Halab (Aleppo)
Al-Lādhiqīyah
Hamāh • SYRIA
NORTH CYPRUS
Nicosia • CYPRUS
Hims
Tudmur (Palmyra)
Dayr az Zawr

M E D I T E R R A N E A N   S E A

Tripoli (Tarābulus)
Al Khums
Misrātah
Zlitan
Qasr Banī Walīd
Al Qaryah
Ash Sharqīyah
Al Mari
Zāwiyat al Baydā • Darnah
Tūkrah
BARQAH (CYRENAICA)
AL JABAL AL AKHDAR
Tubruq
Sīdī Barrānī
As Sallūm
Marsā Matrūh
Al Alamayn
ALEXANDRIA (Al Iskandarīyah)
Dumyāt
Damanhūr • Tanta
Al Mansūrah
Port Said
Ghazzah
Az Zaqāzīq
Suez (As Suways)
SINAI PEN.
Jabal Katrīna

LEBANON
Beirut
Damascus (Dimashq)
Haifa
Tel Aviv-Yafo
ISRAEL
Jerusalem
Amman
JORDAN
IRAQ
SYRIAN DESERT
(BĀDIYAT ASH SHĀM)
Al Jawf
An Nafūd

Khalīj Surt
Surt
An Nawfalīyah
Ajdābiyah
Al Uqaylah
Qasr al Burayqah
ABULUS (TRIPOLITANIA)

JABAL AS SAWDĀ
Sawknah
Zillah
Zaltan
Tarbū
Waw al-Kabīr

AZZĀN (FEZZAN)
IDEHAN
MARZŪQ
Mārzuq
SARĪR TIBASTI

Marādah
Awjilah
Wāhāt Jālū
Al Jaghbūb
Siwah (Oasis)

L I B Y A

E G Y P T

ARABIAN DESERT

L I B Y A N
D E S E R T
(AS SAHRĀ AL LĪBĪYAH)

Al Jaghbūb
MUNKHAFAD AL QATTĀRAH
-436
Al Bawīti
Al Fayyūm
Banī Suwayf
Al Minyā
Asyūt • Akhmīm
Sawhāj
Thebes (Ruins)
Al Uqsur (Luxor)
Idfū
Aswān High Dam • Aswān

Qasr al Farāfirah

Qasr al Farāfirah

Buzaymah
Rebiana (Oasis)
Al Kufrah (Oasis)
Al Jawf

Maʼtan Bishārah
Biʼr Misāhah • Ash Shabb

Qinā • Al Qusayr
Būr Safājah
RAʼS BANĀS
ADMINISTRATIVE BDY.
Halaʼib

S A U D I
A R A B I A
NAJD
AL HIJAZ
Taymāʼ
Hāʼil
Buraydah
Al Wajh
Yanbuʼ
Al Madīnah (Medina)
Jiddah
Mecca (Makkah)
Al Khurmah

RED SEA

Pic Toussidé 10 712
TIBESTI
Emi Koussi 11 204

Ounianga Kébir

NUBIAN DESERT

Arbi • Kosha
Dalqū
3rd Cataract
Dunqulah
Al Khandaq • Kuraymah
Kürtī
Al ʼAtrūn
Ad Dabbah
4th Cataract
Marawi
Abū Hamad
Barbar
ʼAtbarah
Ad Dāmir
Būr Sūdān
Sawākin
Tawkar
Taqatu Hayya
Jabal Erba 7 274

Jiddah
Al Qunfudhah
Abha
JĀZIʼR FARASĀN
DAHLAK ARCH.
KAMARAN

BORKOU
Largeau
BODELE
Agadem (Oasis)
Bilma
Fada
ENNEDI
Oum Chalouba

Yarda

Wadi al Malik

C H A D

Lake Chad
Lac Tchad
Mao
Abéché
OUADDAĪ
Yao
Am Timan
N'Djamena (Fort-Lamy)
MANDARA MTS.
Maroua
Boussou
Léré
Laī
Sarh

S U D A N
KURDUFĀN
Al Fāshir
DĀRFUR
Jabal Marrah 10 131
An Nuhūd
Al Uqayyah
Nyala
Babanūsah
Talawdī
Al Ubayyid
Ad Duwaym
Sannār
Al Qadārif
Kūstī
AN NUBĀH
JIBĀL NUBĀH

5th Cataract
6th Cataract
Wad al Malik
Shandī
Omdurman (Umm Durman)
Al Khartūm Bahrī
Khartoum (Al Khartūm)
Rufāʼah
Al Kāmilīn
Wad Madanī
Sinjah
Qallābāt
Semnar Dam
Roseires Res.
Ar Rusayrīs
Kurmuk
Asosa

Kassalā
Sebderat
Adi Ugri
Om Hajer
 Barentu
Akordat
Keren
Misiwa (Massawa)
Asmera
Mersa Fatma
Adwa
Gondar
Ras Dashen Terara 15 158
DENAKIL
Adi Ugri

E R I T R E A
Al Hudaydah
YEMEN
Al Mukha
Ed
Beylul
Aseb

Debre Tabor
Dangila
Debre Markos
13 494
14 478
Addis Ababa (Adis Abeba)
Dire Dawa
Harer
CHINĀMI MTS.
HARERGE

CENTRAL AFRICAN REPUBLIC
Koundé
Bouar
Fort-Sibut
Fort-de-Possel
Bambari
CHAĪNE DES MONGOS
Ndélé
Ouanda Djallé
Yalinga
Rafaī
Zémio
Gwane
Bangassou
Mobaye
Bondo
Bambesa
BAHR AL GHAZĀL
Kafia Kingi
Waw
Mashraʼar Raqq
Rumbek
Tambura
AS SUDD
Shambe
Bor
Juba
Mongalla
Kapoeta

Jima
Goba
Ginir
Soda
Werta
Welo
SIDAMO
Mega
Moyale
El Wak
Doolow

DEMOCRATIC REPUBLIC OF THE CONGO
CONGO
Bangui
Mbaïki
Libenge
Zongo
Mobayi-Mbongo
Gemena
Businga
Aketi
Buta
Lisala
Bumba
Panga
Basoko
Kisangani (Stanleyville)
Boyoma Falls
Isangi
Mbandaka
Dongou
Impfondo
Bomongo
Basankusu
Makanza

Yokaduma
Lomié
Ngaoundéré
Kounde
Carnot
Berbérati
Nola
Ouesso

UGANDA
Mahagi Port
Arua
Nimule
Kitgum
Ft. Portal
Margherita Peak 16 763
Kampala
Entebbe
Jinja
Lake Victoria

K E N Y A
Soroti
Eldoret
Meru

S O M A L I A

Bumba
Avakubi
Niangara
Watsa
Gombari
Isiro
Dungu
Faradje
Aba
Masindi
Trumu

Equator

ETHIOPIA
Dembi Dolo
Gambela
Gore
Nasir
Malakāl
Kodok
Lol
Shewa Gimira
Maji
Bako
Mek.
Mékèmte
Fiche
Taju Woj
Admin. Bdy.
Lake Stefanie

0   50   100   200   300   400   500 Miles
0   100   200   400   600   800 Kilometers

5° · 10° · 15° · 20° · 25° · 30° · 35° · 40°

Scale 1:16 000 000; one inch to 250 miles. Sinusoidal Projection
Elevations and depressions are given in feet

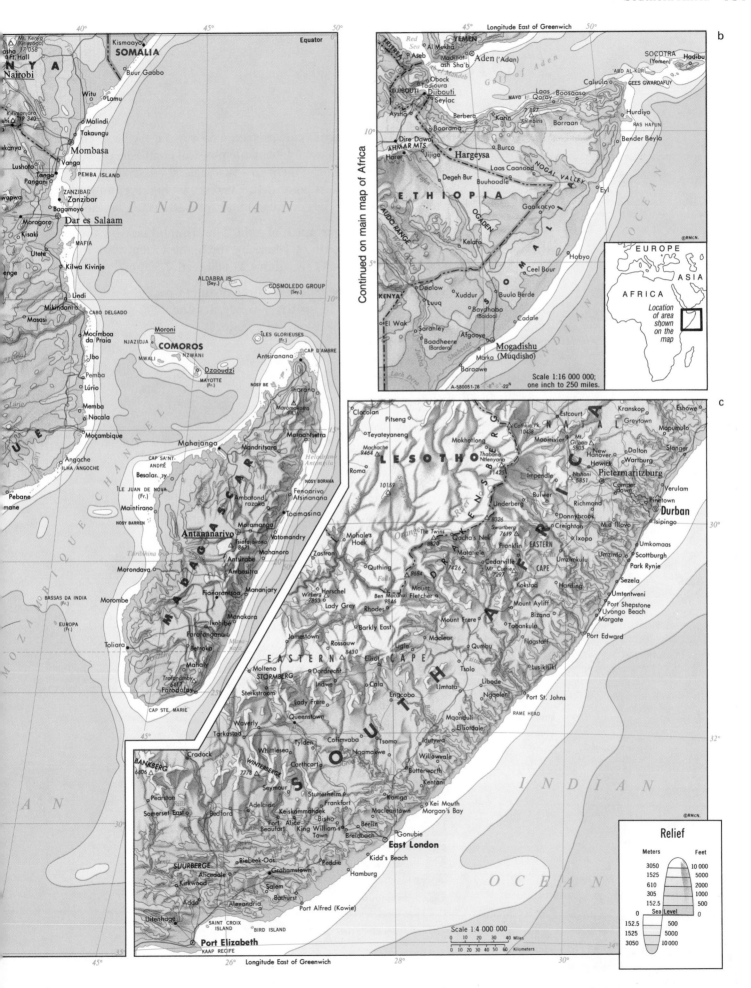

b

Continued on main map of Africa

Longitude East of Greenwich

YEMEN
Red Sea
Al Mukha
ERITREA
Madinat ash Sha'b
Aden ('Adan)
Aseb
Gulf of Aden
SOCOTRA (Yemen)
Hadibu
'ABD AL-KURI
GEES GWARDAFUY
Obock
Tadjoura
DJIBOUTI
Djibouti
Seylac
Berbera
Caluula
Boosaaso
RAS HAFUN
Aysha
Boorama
Karin
Shimbiris 897
Borraan
Hurdiyo
Bender Beyla
Dire Dawa
AHMAR MTS.
Harer
Jijiga
Hargeysa
Laas Caanood
NOGAL VALLEY
Eyl
ETHIOPIA
Degeh Bur
Buuhoodle
Gaalkacyo
AUDO RANGE
OGADEN
Kelafo
Hobyo
Ceel Buur
KENYA
Doolow
Xuddur
Buulo Berde
El Wak
Luuq
Baydhabo (Baidoa)
Cadale
Saranley
Afgooye
Baadheere (Barderal)
Marka (Muqdisho)
Mogadishu
Shabeelle
Baraawe
Lach Dera
A-580051-76

EUROPE
ASIA
AFRICA
Location of area shown on the map

Scale 1:16 000 000;
one inch to 250 miles.

c

N Y A
Kismaayo
SOMALIA
Equator
Mt. Kenya (Kirinyaga) 17,058
asha
Ft. Hall
Nairobi
Buur Gaabo
Witu
Lamu
Kilimanjaro 19,340
kànya
Malindi
Takaungu
Mombasa
Lushoto
Vanga
Tanga
PEMBA ISLAND
Pangani
wapwa
ZANZIBAR
Zanzibar
INDIAN
Morogoro
Bagamoyo
Kisaki
Dar es Salaam
Utete
MAFIA
enge
Kilwa Kivinje
Lindi
Mikindani
ALDABRA IS. (Sey.)
COSMOLEDO GROUP (Sey.)
Masasi
CABO DELGADO
Moçimboa da Praia
Moroni
Ibo
NJAZIDJA
COMOROS
ÎLES GLORIEUSES (Fr.)
Pemba
MWALI
NZWANI
Lúrio
Memba
Dzaoudzi
MAYOTTE (Fr.)
Antsiranana
CAP D'AMBRE
Nacala
Moçambique
NOSY BE
Iharana
UE
Angoche
ILHA ANGOCHE
Mahajanga
Maromokotro
Maroantsetra
Pebane
mane
CAP SAINT-ANDRÉ
Helodranon' Antongila
NOSY BORAHA
Mandritsara
Besalampy
Ambatondrazaka
Fenoarivo Afsinanana
Clocolan
Pitseng
Estcourt
NATAL
Kranskop
Eshowe
Greytown
Mapumulo
Teyateyaneng
Mokhotlong
Camino Pk. 10438
Mooirivier
Mt. Gilboa 6361
Dalton
Stanger
Machache 9464
LESOTHO
Thabana Ntlenyana
New Hanover
Howick
ÎLE JUAN DE NOVA (Fr.)
Maintirano
Toamasina
11425
Impendle
Ntshoni 5851
Pietermaritzburg
Verulam
NOSY BARREN
Moramanga
Roma
10159
Underberg
Bulwer
Richmond
Camperdown
Pinetown
Antananarivo
Tsiafajavona 8671
Vatomandry
8326
Donnybrook
Mid Illovo
Durban
Isipingo
Morondava
Antsirabe
Mahanoro
Mohale's Hoek
The Twins
Swartberg 7619
Gacha's Nek
Creighton
Ixopo
Umkomaas
BASSAS DA INDIA (Fr.)
Ambositra
Mananjary
Zastron
Matatiele
EASTERN
Umzinto
Scottburgh
Morombe
Fianarantsoa
Quthing
9684
Franklin
CAPE
Park Rynie
EUROPA (Fr.)
Manakara
7426
Cedarville
Kokstad
Harding
Sezela
Ivohibé
Mount Fletcher
Mt. Currie 7297
Umtentweni
Toliara
Farafangana
Witberg 2853
Herschel
Ben Macdhui 9846
Bizana
Port Shepstone
Uvongo Beach
Betroka
Lady Grey
Rhodes
Mount Frere
Tabankulu
Margate
Mahaly
Jamestown
Rossouw
Mount Aylff
Flagstaff
Port Edward
Tsaratanana 4317
Ambovombe
Barkly East
Elliot
8430
Maclear
Qumbu
Lusikisiki
CAP STE. MARIE
Molteno
Dordrecht
Cala
Tsolo
Libode
Port St. Johns
STORMBERG
Indwe
Engcobo
Umtata
Ngqeleni
RAME HEAD
Waverly
Sterkstroom
Ugie
Mqanduli
Elliotdale
Tarkastad
Queenstown
Tsomo
Idutywa
Cradock
Tylden
Cofimvaba
Ngamakwe
Willowvale
BANKBERG 6606
Whittlesea
Carthcart
SOUTH
Butterworth
WINTERBERG 7778
Seymour
Stutterheim
Komga
Kentani
Adelaide
Frankfort
Macleantown
Kei Mouth
Morgan's Bay
Somerset East
Bedford
Keiskammahoek
Bisho
Berlin
Fort Beaufort
Fort Alice
King William's Town
Breidbach
Gonubie
East London
SUURBERGE
Riebeek-Oos
Peddie
Hamburg
Alicedale
Grahamstown
Kidd's Beach
Kirkwood
Salem
Bathurst
Addo
Alexandria
Port Alfred (Kowie)
Uitenhage
SAINT CROIX ISLAND
BIRD ISLAND
Port Elizabeth
KAAP RECIFE

Scale 1:4 000 000
0 10 20 30 40 Miles
0 10 20 30 40 50 60 Kilometers

Longitude East of Greenwich

Relief

Meters		Feet
3050		10 000
1525		5000
610		2000
305		1000
152.5		500
0	Sea Level	0
152.5		500
1525		5000
3050		10 000

INDIAN OCEAN
MOZAMBIQUE CHANNEL
MADAGASCAR

Asia

Covering nearly one-third of the Earth's land surface, Asia is by far the largest of the seven continents. It is a land of extremes and dramatic physical contrasts, containing nearly every type of landform, and many of them on a vast scale. It boasts the world's lowest point (the Dead Sea), its highest point (Mt. Everest), its highest and largest plateau (the Plateau of Tibet), and its largest inland body of water (the Caspian Sea).

Wide belts of mountain systems cover much of Asia. The Himalayas, which form a great 1,500-mile (2,400-km) arc south of Tibet, are the highest mountains in the world: more than 90 Himalayan peaks rise above 24,000 feet (7,320 m).

The beginnings of civilization can be traced to three distinct areas of Asia: Mesopotamia, around 4000 BC; the Indus River valley, around 3000 BC; and China, around 2000 BC. Eight of the world's major religions—Buddhism, Christianity, Confucianism, Hinduism, Islam, Judaism, Shinto, and Taoism—originated in Asia.

Asia at a glance

Land area:
17,300,000 square miles
(44,900,000 sq km)

Estimated population:
3,761,165,000

Population density:
217/square mile (84/sq km)

Mean elevation: 3,000 feet
(910 m)

Highest point: Mt. Everest,
China (Tibet)-Nepal, 29,028
feet (8,848 m)

Lowest point: Dead Sea,
Israel-Jordan, 1,339 feet
(408 m) below sea level

Longest river: Yangtze
(Chang), 3,900 mi (6,300 km)

Number of countries
(incl. dependencies): 50

Largest independent country:
Russia (Europe/Asia),
6,592,849 square miles
(17,075,400 sq km)

Smallest independent country:
Maldives, 115 square miles
(298 sq km)

Most populous independent
country: China, 1,278,720,000

Least populous independent
country: Maldives, 315,000

Largest city: Mumbai, India,
pop. 11,914,398

Annapurna, one of the highest mountains in the Himalayas

Coldest place:
Verkhoyansk, Russia
-90∞F (-68∞C)

Lowest point:
Dead Sea, Israel-Jordan
1,339 ft (408 m) below sea level

Hottest place:
Tirat Zvi, Israel
129∞F (54∞C)

Wettest place:
Cherrapunji, India
450 inches (1143 cm)/year

Driest place:
Aden, Yemen
1.8 inches (4.6 cm)/year

Highest point:
Mt. Everest, China (Tibet)-Nepal
29,028 ft (8,848 m)

URALS · Ob' · Western Siberian Lowland · Yenisey · Siberia · Arctic Circle · Sea of Okhotsk · Poluostrov Kamchatka · Kirghiz Steppe · CAUCASUS · ALTAI MTS. · Amur · Sea of Japan · HONSHU · ZAGROS MTS. · Plateau of Iran · PAMIRS · TIEN SHAN · Gobi Desert · Huang · East China Sea · Pacific Ocean · KUNLUN SHAN · Plateau of Tibet · Yangtze · Arabian Peninsula · Indus · HIMALAYAS · Ganges · Tropic of Cancer · LUZON · Arabian Sea · Deccan · Bay of Bengal · South China Sea · Malay Pen. · SUMATRA · BORNEO · Equator · NEW GUINEA

Landforms

- Mountains
- Widely spaced mountains
- High tablelands
- Hills and low tablelands
- Plains
- Depresssions, basins
- High tablelands and ice caps
- Mountains and ice caps

© Rand McNally & Co.
N-ANS60000-A3- - - -2

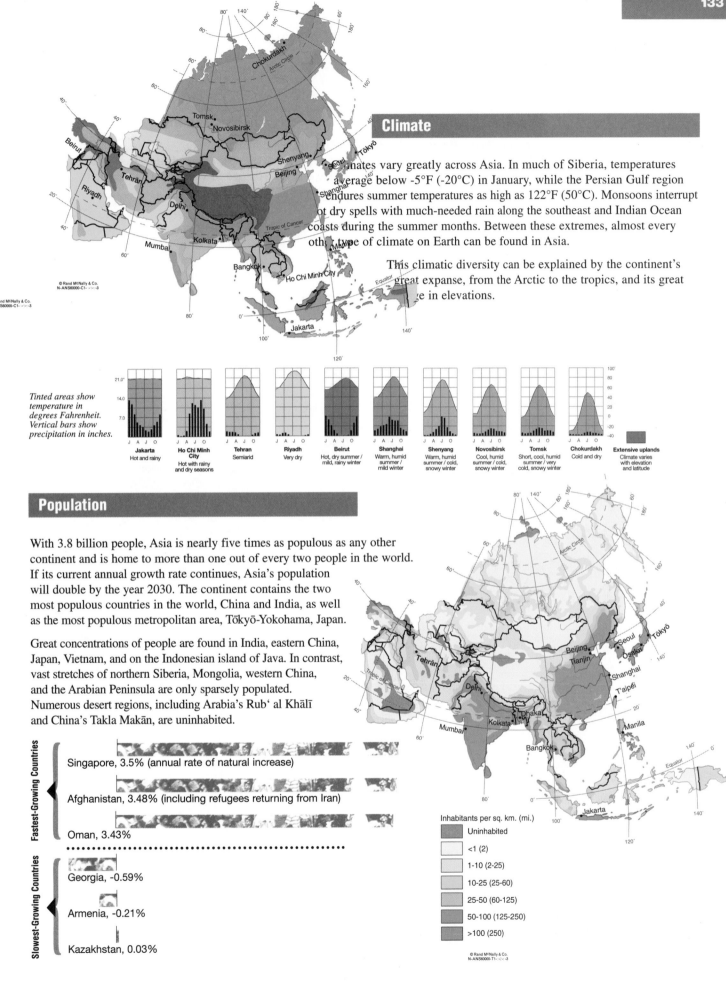

Climate

Climates vary greatly across Asia. In much of Siberia, temperatures average below -5°F (-20°C) in January, while the Persian Gulf region endures summer temperatures as high as 122°F (50°C). Monsoons interrupt hot dry spells with much-needed rain along the southeast and Indian Ocean coasts during the summer months. Between these extremes, almost every other type of climate on Earth can be found in Asia.

This climatic diversity can be explained by the continent's great expanse, from the Arctic to the tropics, and its great range in elevations.

Tinted areas show temperature in degrees Fahrenheit. Vertical bars show precipitation in inches.

Jakarta Hot and rainy	**Ho Chi Minh City** Hot with rainy and dry seasons	**Tehran** Semiarid
Riyadh Very dry	**Beirut** Hot, dry summer / mild, rainy winter	**Shanghai** Warm, humid summer / mild winter
Shenyang Warm, humid summer / cold, snowy winter	**Novosibirsk** Cool, humid summer / cold, snowy winter	**Tomsk** Short, cool, humid summer / very cold, snowy winter
Chokurdakh Cold and dry	**Extensive uplands** Climate varies with elevation and latitude	

Population

With 3.8 billion people, Asia is nearly five times as populous as any other continent and is home to more than one out of every two people in the world. If its current annual growth rate continues, Asia's population will double by the year 2030. The continent contains the two most populous countries in the world, China and India, as well as the most populous metropolitan area, Tōkyō-Yokohama, Japan.

Great concentrations of people are found in India, eastern China, Japan, Vietnam, and on the Indonesian island of Java. In contrast, vast stretches of northern Siberia, Mongolia, western China, and the Arabian Peninsula are only sparsely populated. Numerous desert regions, including Arabia's Rub' al Khālī and China's Takla Makān, are uninhabited.

Fastest-Growing Countries

Singapore, 3.5% (annual rate of natural increase)

Afghanistan, 3.48% (including refugees returning from Iran)

Oman, 3.43%

Slowest-Growing Countries

Georgia, -0.59%

Armenia, -0.21%

Kazakhstan, 0.03%

Inhabitants per sq. km. (mi.)

- Uninhabited
- <1 (2)
- 1-10 (2-25)
- 10-25 (25-60)
- 25-50 (60-125)
- 50-100 (125-250)
- >100 (250)

© Rand McNally & Co.
N-ANS60000-T1- -:-:- -3

© Rand McNally & Co.
N-ANS60000-C1- -:-:- -3

Environments and Land Use

Despite rapid industrialization in Japan, Korea, and Singapore, feeding the enormous and fast-growing population remains Asia's primary economic focus. In China, India, and Indonesia, two-thirds of the work force is engaged in farming. Where arable land exists, it is generally cultivated intensively. Rice is the most commonly grown crop: the continent produces more than 90% of the world's total. Other important crops include wheat, sorghum, millet, maize, and barley.

Asia's major agricultural regions are found in the fertile alluvial valleys, floodplains, and deltas of some of its greatest rivers, such as the Ganges and Brahmaputra in northern India, the Indus in Pakistan, the Huang (Yellow) and Yangtze in eastern China, the Irawaddy in Myanmar (Burma), the Mekong in Cambodia and Vietnam, and the Tigris and Euphrates in Iraq.

Tundra vegetation prevails across the arctic and subarctic regions of Siberia. Farther south, much of the land is densely forested. Deforestation, however, is rampant across the continent. In the colder central and northern areas, whole forests have been cut down to provide wood for heat and cooking. The tropical rain forests of the Indochina Peninsula, Malaysia, Indonesia, and the Philippines are rapidly being destroyed for their valuable hardwood, especially teak.

A wide sweep of semiarid grasslands across Central Asia covers one-quarter of the continent. These immense grazing lands are used by the people of many countries—notably Kazakhstan and Mongolia—for livestock that include almost one-third of the world's cattle, nearly three-fifths of its goats, and half of its pigs. In recent decades, the tremendous petroleum reserves located in the arid west, around the Persian Gulf, have been both a source of wealth and a cause of turmoil. The continent also has less exploited, but sizable reserves of natural gas in Siberia and coal in China.

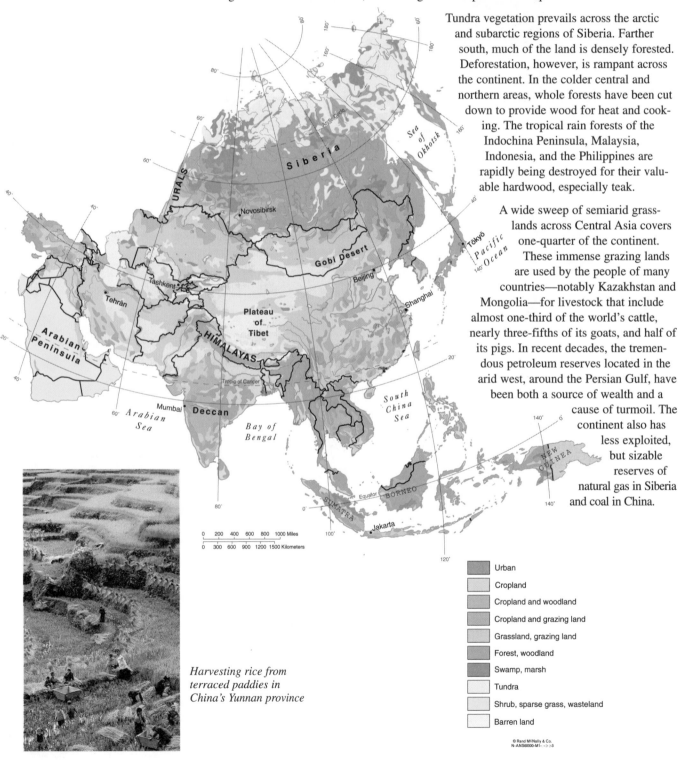

Harvesting rice from terraced paddies in China's Yunnan province

Urban
Cropland
Cropland and woodland
Cropland and grazing land
Grassland, grazing land
Forest, woodland
Swamp, marsh
Tundra
Shrub, sparse grass, wasteland
Barren land

Asia and the Ring of Fire

"Ring of Fire" is the dramatic name given to the volcanoes that nearly encircle the Pacific Ocean. Its existence is explained by plate tectonic theory. According to the theory, Earth's thin crust is broken into sections or plates that move relative to each other. (See Plate Tectonics, pages 12 and 13.)

The giant Pacific plate submerges beneath other plates and plunges into Earth's hot interior, creating deep ocean trenches and causing magma to rise to the surface and erupt as volcanoes. The moving plates also trigger thousands of earthquakes that rumble along the Ring of Fire every year.

The Asian section of the Ring of Fire starts in the north on the Kamchatka Peninsula; swings down through the Kuril Islands; and includes Japan, Taiwan, the Philippines, and Indonesia.

The Kamchatka Peninsula, about the size of California, has more than 100 volcanoes. About 20 of these are active. Klyuchevskaya, the largest active volcano in Kamchatka, and in Eurasia, puts out about 60 million tons of basalt a year.

Japan is one of the most earthquake prone regions in the world. Four plates come together at Japan, and their movement against each other creates constant strain on the earth's crust. In 1995, an earthquake at Kōbe resulted in 5,500 fatalities. Japan's most deadly quake was in 1923, when more than 140,000 lives were lost.

The largest volcanic eruption in recorded history occurred in Indonesia. In 1815, Mt. Tambora spewed a column of ash 28 miles into the air. The fallout ruined crops, causing widespread famine. 82,000 people died as a result of Tambora's eruption, most of them from starvation. Gases from Tambora circled the globe and lowered world temperatures by as much as 3°. In 1816, the "year without summer", killing frosts during summer months destroyed crops in Europe and North America.

The 1991 eruption of Mt. Pinatubo in the Philippines was the second largest volcanic eruption of the 20th century, and approximately ten times as large as Mt. St. Helen's. Pinatubo blew an ash column about 21 miles into the sky. Although Pinatubo is near a heavily populated area not far from Manila, authorities were able to warn and evacuate residents in advance, preventing what could have been a major disaster.

Mt. Pinatubo, Philippines

© Rand McNally & Co.
N-ANS60000-B3- - -·-·-1

0   200   400   600   800   1000 Miles

0   300   600   900   1200 1500 Kilometers

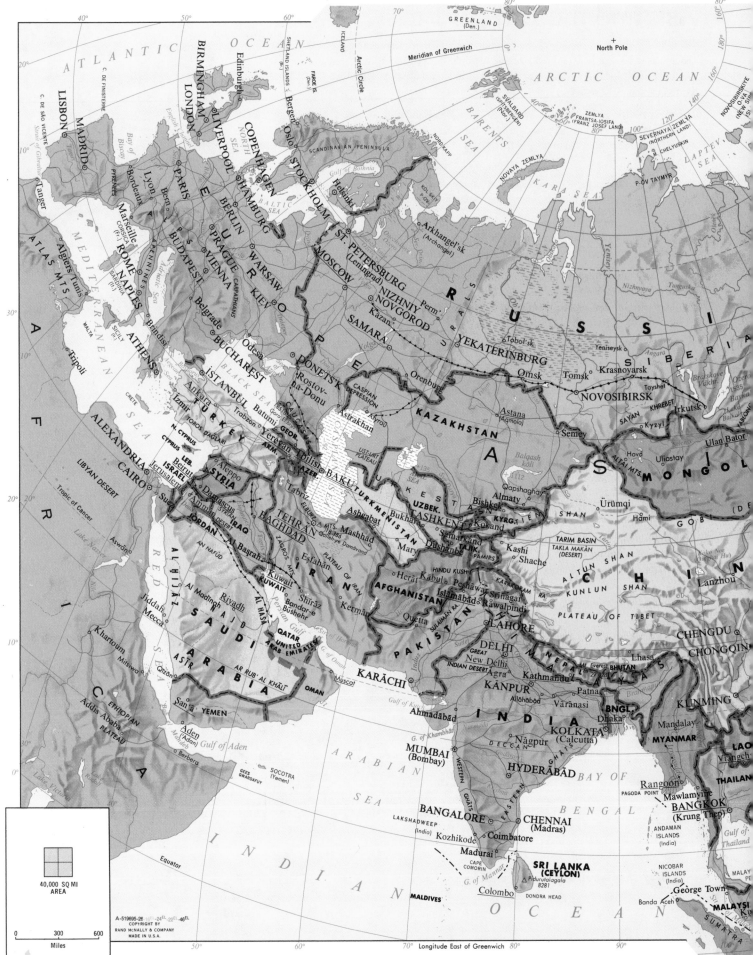

Scale 1:40 000 000; one inch to 630 miles. Lambert's Azimuthal, Equal Area Projection
Elevations and depressions are given in feet

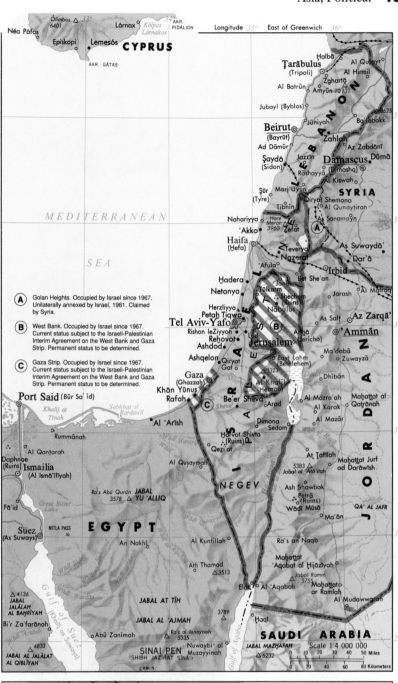

Scale 1:40 000 000; one inch to 630 miles. Lambert's Azimuthal, Equal Area Projection
Elevations and depressions are given in feet

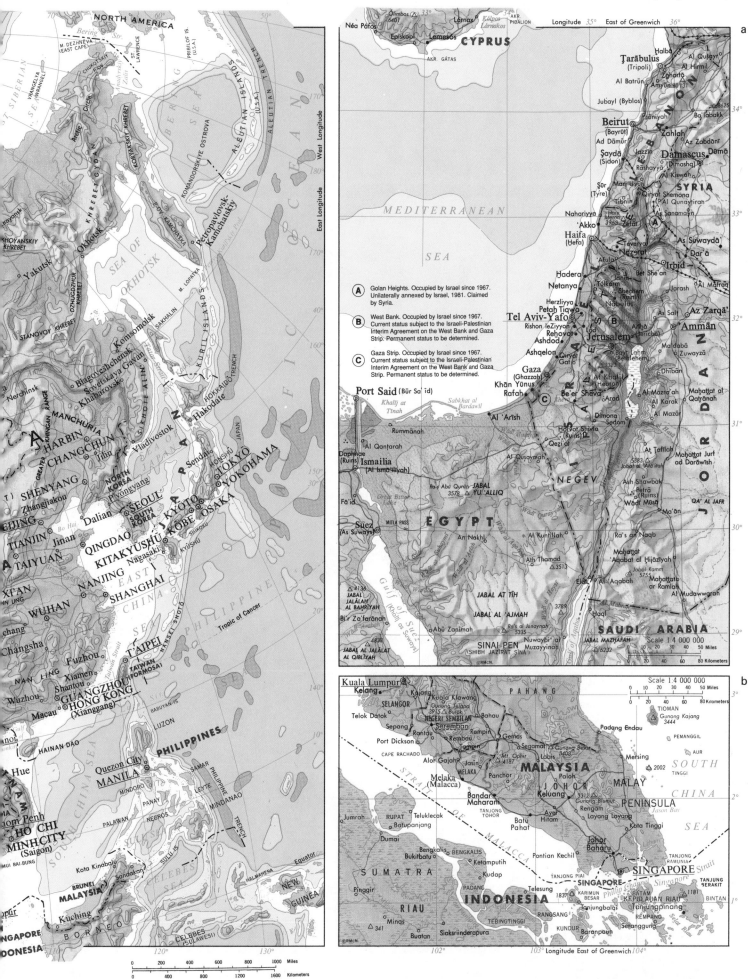

Cities and Towns

0 to 50,000 ○
50,000 to 500,000 ⊙
500,000 to 1,000,000 ◎
1,000,000 and over

Scale 1:16 000 000; one inch to 250 miles Conic Projection
Elevations and depressions are given in feet.

Relief

Meters		Feet
3050		10 000
1525		5000
610		2000
305		1000
152.5		500
0	Sea Level	0
152.5		500
1525		5000
3050		10 000

BLACK SEA

TURKEY
SYRIA
IRAQ
IRAN
SAUDI
ARABIA
OMAN
YEMEN
EGYPT
SUDAN
SOMALIA
ERITREA
ETHIOPIA
JORDAN
ISRAEL
LEBANON
NORTH CYPRUS
CYPRUS
KUWAIT
QATAR
BAHRAIN
UNITED ARAB EMIRATES
TURKMENISTAN
UZBEKISTAN
KAZA
RUSSIA
GEORGIA
ARMENIA
AZERBAIJAN
AFGHA
KURDISTAN

MEDITERRANEAN SEA
CASPIAN SEA
PERSIAN GULF
GULF OF OMAN
GULF OF ADEN
GULF OF SUEZ
GULF OF AQABA
ARABIAN SEA

AN NAFŪD
AD DAHNA
AR RUB' AL KHĀLĪ
DASHT-E KAVIR DESERT
PLATEAU OF IRAN
SYRIAN DESERT
KARA-KUM (DESERT)
ELBURZ MTS.
CAUCASUS
TOROS DAĞLARI

ISTANBUL, Izmir, Bursa, Ankara, Konya, Adana, Aleppo, Damascus, Beirut, Tel Aviv-Yafo, Jerusalem, Gaza, CAIRO (Al Qāhirah), ALEXANDRIA (Al Iskandarīyah), Port Said, Suez, Amman, BAGHDAD, Karbalā', An Najaf, Al Başrah, KUWAIT (Al Kuwayt), Abādān, Ahvāz, Shīrāz, Eşfahān, TEHRAN, Hamadān, Bakhtarān, Mashhad, Ashgabat, RIYADH (Ar Riyād), Mecca (Makkah), Al Madīnah (Medina), Jiddah, SAN'Ā, Aden ('Adan), Al Hudaydah, Al Mukallā, Muscat, Ad Dawhah, Al Manāmah, Ad Dammām, BAHRAIN, Abū Zaby, Dubayy

Tropic of Cancer

Longitude East of Greenwich

Relief

Meters		Feet
3050		10 000
1525		5000
610		2000
305		1000
152.5		500
0	Sea Level	0
152.5		500
1525		5000
3050		10 000

Below Sea Level

A-569400-76 11-24-21-43
COPYRIGHT BY
RAND McNALLY & COMPANY
MADE IN U.S.A.

Scale 1:16 000 000; one inch to 250 miles. Polyconic Projection
Elevations and depressions are given in feet

Areas occupied by Israel since 1967

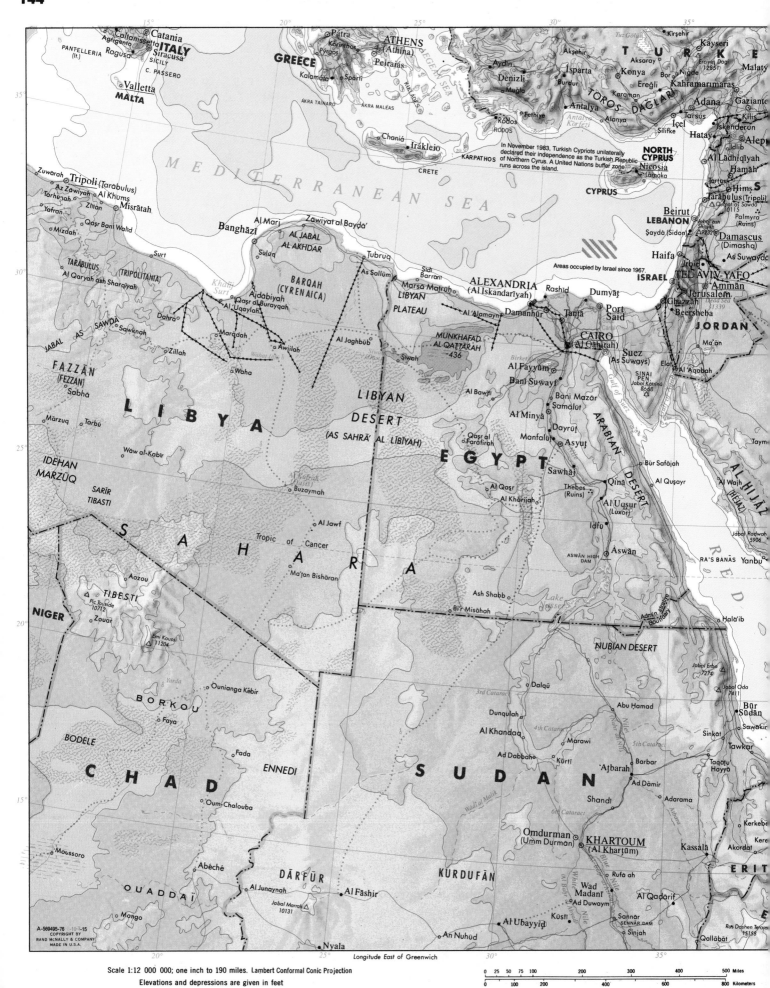

MEDITERRANEAN SEA

ITALY
Catania
Agrigento
Caltanissetta
Ragusa
SICILY
Siracusa
C. PASSERO
PANTELLERIA
(It.)
MALTA
Valletta

GREECE
Pátra
Kórinthos
Pýrgos
Kalamáta
Spárti
ÁKRA TAÍNARO
ÁKRA MALÉAS
ATHENS
(Athína)
Peiraiás
RÓDOS
Chaniá
Irákleio
CRETE
KÁRPATHOS
KARPATHOS
RÓDOS

TURKE
Tuz Gölü
Kirşehir
Kayseri
Aydin
Akşehir
Aksaray
Bor
Niğde
Ercives Daği
12851
Malaty
Denizli
Isparta
Burdur
Konya
Ereğli
Karaman
Kahramanmaraş
Muğla
Fethiye
Antalya
Alanya
TOROS DAĞLARI
Silifke
Içel
Tarsus
Adana
Gazian
Antalya
Körfezi
Kilis
İskenderun
Hatay
Ale
Idlib
Al Ladhiqiyah
Hamāh
Tartūs
NORTH CYPRUS
Nicosia
Lárnaka
CYPRUS
In November 1983, Turkish Cypriots unilaterally declared their independence as the Turkish Republic of Northern Cyrus. A United Nations buffer zone runs across the island.
Qal'at as Sawdā'
1115
Palmyra (Ruins)
Beirut
LEBANON
Şaydā (Sidon)
Jabal ash Shaykh 9232
Damascus
(Dimashq)
As Suwaydā'
Haifa
Irbid
Areas occupied by Israel since 1967.
ISRAEL
TEL AVIV-YAFO
Amman
Jerusalem
Ghazzah
Dead Sea 1339
Beersheba
JORDAN
Ma'an
Elat
Al 'Aqabah

Zuwārah
Tripoli (Tarābulus)
Az Zāwiyah
Al Khums
Misrātah
Tārhunah
Zlitan
Yafran
Qaşr Bani Walid
Mizdah
Banghāzī
Al Marj
Zāwiyat al Baydā'
AL JABAL AL AKHDAR
Sulūq
Surt
Tubruq
As Sallūm
Sidi Barrāni
Marsa Maţrūh
ALEXANDRIA (Al Iskandarīyah)
Rashid
Dumyāţ
Port Said
TARABULUS (TRIPOLITANIA)
Al Qaryah ash Sharqiyah
BARQAH (CYRENAICA)
Ajdābiyah
Qaşr al Burayqah
Al 'Uqaylah
Dahra
LIBYAN PLATEAU
Al 'Alamayn
Damanhūr
Tanta
Khalīj Surt
Al Jaghbūb
MUNKHAFAD AL QATTĀRAH -436
Siwah
Wāhāt Sīwah
CAIRO (Al Qāhirah)
Suez (As Suways)
JABAL AS SAWDĀ'
Sawknah
Marādah
Awjilah
Al Fayyūm
Birkat Qārūn
Bani Suwayf
SINAI PEN.
Jabal Katrina 8668
FAZZĀN (FEZZAN)
Zillah
Wāha
Al Bawīţī
Bani Mazār
Samalūţ
Sabhā
Al Minyā
Dayrūţ
ARABIAN DESERT
Al Wajh (HEJAZ)
Mārzuq
Tarbū
Waw al-Kabir
LIBYA
Qaşr al Farāfirah
Manfalūţ
Asyūţ
EGYPT
Al Qaşr
Sawhaj
Būr Safājah
AL HIJAZ (HEJAZ)
IDEHAN MARZŪQ
SARĪR TIBASTI
Al Khārijah
Thebes (Ruins)
Qinā
Al Qusayr
Taym
Buzaymah
Al Khārijah
Al Uqşur (Luxor)
Idfu
Jabal Radwah 5906
SAHARA
Al Jawf
Tropic of Cancer
Aswān High Dam
Aswān
RA'S BANĀS
Aozou
Ma'tan Bishāran
Ash Shabb
Bi'r Misāhah
Lake Nasser
Halā'ib
TIBESTI
Pic Toussidé 10712
Zouar
Emi Koussi 11204
NUBIAN DESERT
Jabal Erba 7274
NIGER
Admin. Boundary
Jabal Oda 7411
Ouinianga Kébir
Dalqū
BORKOU
Faya
3rd Cataract
Dunqulah
Abu Hamad
Būr Sūdān
BODELE
Al Khandaq
4th Cataract
Marawi
Sawakin
Fada
Ad Dabbah
Kūrti
Barbar
Sinkāt
ENNEDI
SUDAN
Atbarah
Taqatu Hayyā
Moussoro
Oum-Chalouba
5th Cataract
Ad Dāmir
Tawkar
CHAD
Shandi
Adarama
Abéché
DĀRFŪR
KURDUFĀN
6th Cataract
Kerkebe
Oum-Hadjer
Al Junaynah
Jabal Marrah 10131
Al Fāshir
Omdurman (Umm Durmān)
KHARTOUM (Al Khartūm)
Kassalā
Akordat
Mongo
OUADDAÏ
Wad Madanī
Ad Duwaym
Al Qadārif
ERI
Abéché
Kūstī
Rufa'ah
Sannar
Sennār Dam
Ras Dashen Terara 15158
Nyala
An Nuhūd
Al Ubayyid
Sinjah
Qallābāt

Longitude East of Greenwich

Scale 1:12 000 000; one inch to 190 miles. Lambert Conformal Conic Projection
Elevations and depressions are given in feet

0 25 50 75 100   200   300   400   500 Miles
0   100   200   400   600   800 Kilometers

ARMENIA
AZERBAIJAN
BAKU
(Bakı)

Erzurum
Yerevan
AZER.
Naxçıvan
Xankändi
(Stepanakert)
Mt. Ararat
16854

Turkmenbashy
Celeken
Nebitdag
Gyzylarbat

TURKMENISTAN

60°   65°

35°

Mus
Tatvan
Bitlis
Van
Khvoy
Marand
Ahar
Astara
Länkäran
Salyan
CASPIAN
SEA
Surface 92 Feet Below Sea Level
Gyzyletrek a
Bojnurd
Ashgabat
KOPPEH
Quchan
DAGH
Saragt
Mary
Yöloten
Andkhvoy
Meymaneh

Elâziğ
Diyarbakir
Siverek
nlıurfa
Mardin
KURDISTAN
Orümiyeh
Tabrīz
Ardabil
Maragheh
Mianeh
Rasht
Lähijän
Bandar-e Anzali
Bandar-e
Torkeman
Gorgan
Babol
Chālūs
Emamshahr
Sabzevär
Neyshäbür
Binalud
11208
Mashhad
Torbat-e
Heydarīyeh
Torbat-e Jäm
Ghorían o
Herät
Shindand

r az Zawr
Zakho
Al Mawsil
Arbīl
Mahäbäd
Saqqez
Zanjän
Qazvin
Darÿächeh-ye
Oruniyeh
ELBURZ   MTS
Qolleh-ye
Damavand
18386
Rey
Bajestän
Ferdows
Qäyen
Kashmar
Farah

KURD
As Sulaymänīyah
Sanandaj
TEHRAN
DASHT-E KAVIR
DESERT
IRAN
Birjand
AFGHANISTAN

Karkük
Bayjī
Tikrit
Samarra
Hamadän
Qom
Darÿächeh-ye
Namak
Soveh
Rey
PLATEAU OF IRAN
DASHT-E-LÜT
(DESERT)
Nehbandän
Zaranj

Abu Kamal
BAGHDAD
Babylon (Ruins)
Bakhtarän
Arak
Borüjerd
Khorramäbäd
Kashän
Na'in
Esfahän
Yazd
Sürmaq
Darÿächeh-ye
Shähdäd
Namakzär-e
Shähdäd
Chär Borjak
Gowd-e
Zereh

Ar Ramädi
Hadithah
Karbala
Al Küt
Dezfül
Shüshtar
Masjed Soleymän
Häji Gel
Qomsheh
Kerman
Rafsanjan
Zähedän
Lädiz
CHAGAI HILLS
PAKISTAN

SYRIAN
DESERT
IRAQ
MESOPOTAMIA
An Najaf
As Samäwah
An Näsiriyah
Al 'Amärah
Ahväz
Behbehän
Kalür
14100
Persepolis
(Ruins)
Hamún-e
Mäshkel

Badanah
Al Basrah
Khorramshahr
Bandar-e Khomeyni
Gachsärän
Shiräz
Darÿächeh-ye
Bakhtegän
Jahrom
Firgun
10760
Bampür
Chagai Hills

Sakäkah
Rafha
Abädän
KUWAIT
Kazerün
Lär
Bandar-e 'Abbäs
Jäsk
Bandar Beheshti
Gwadar

l Jawf
Al Qoysümah
Kuwait
(Al Kuwayt)
Bandar-e
Büshehr
PERSIAN
GULF
Bandar-e Lengeh
OMAN
Hormuz
GULF   OF   OMAN

AN NAFÜD
Ha'il
JABAL SHAMMAR
Buraydah
Unayzah
AD
DAHNÄ
AL HASÄ
Al Qatif
Ra's at Tannürah
Ad Dammäm
Az Zahrän
(Dhahran)
Dukhän
BAHRAIN
Al Manämah
QATAR
Ad Dawhah
Ash Shäriqah
Dubayy
Abu Zaby
Al Khäbürah
Muscat

diyah
Ash Shaqra
Al Hufuf
UNITED ARAB EMIRATES
AL JABAL
AL AKHDAR
Jabal ash Shäm
9957
RA'S AL HADD
Sür

SAUDI
NAJD
AL AFLAJ
Riyadh
(Ar Riyäd)
As Sulaymänīyah
AD DAHY

Al Madīnah
(Medina)
Mahd adh
Dhahab
Al Mubarraz
Al 'Ubaylah
OMAN

Rabigh
ARABIA
NAFÜD
JABAL TUWAYQ
Al Lidäm
AR   RUB'   AL   KHÄLĪ
RA'S AL MADRAKAH
Al Jawärah
AL MASĪRAH

Jiddah
Mecca (Makkah)
At Tä'if
Al Lith
Qal'at Bishah
NAJRAN
Mirbät

ASIR
Abhä
Al Qunfudhah
Sa'dah
RAMLAT AS
SAB'ATAYN
Shibäm
Say'ün
RA'S FARTAK
Al Ghaydah
ARABIAN

KASR
JĀZA'IR
FARASĀN
Qizan
Al Lubayyah
Sän'a
YEMEN
HADRAMAWT
Sayhut
Ash Shihr
Al Mukalla
SEA

Mitsiwa
DAHLAK
ARCH.
Asmera
KAMARÄN
Ibb
Ta'izz
Shuqrah
Al Hawrah
Hadibu
SUQUTRÄ (SOCOTRA)
(Yemen)

Mekele
DENAKIL
Ramlu
6788
Al Hudaydah
Al Makhä
(Mocha)
Bab el Mandeb
Ash Shi'b
ETHIOPIA
DJIBOUTI
Obock
Aden ('Adan)
Madinat ash Sha'b
GULF   OF   ADEN
Caluula
GEES GWARDAFUY

igrai
Aseb
Seylac
Tadjoura
Djibouti
Qandala
SOMALIA

Relief		
Meters		Feet
3050		10 000
1525		5000
610		2000
305		1000
152.5		500
0	Sea Level	0
		Below
152.5		500 Sea Level
1525		5 000
3050		10 000
6100		20 000

15°

20°

25°

30°

Scale 1:16 000 000; one inch to 250 miles. Polyconic Projection
Elevations and depressions are given in feet

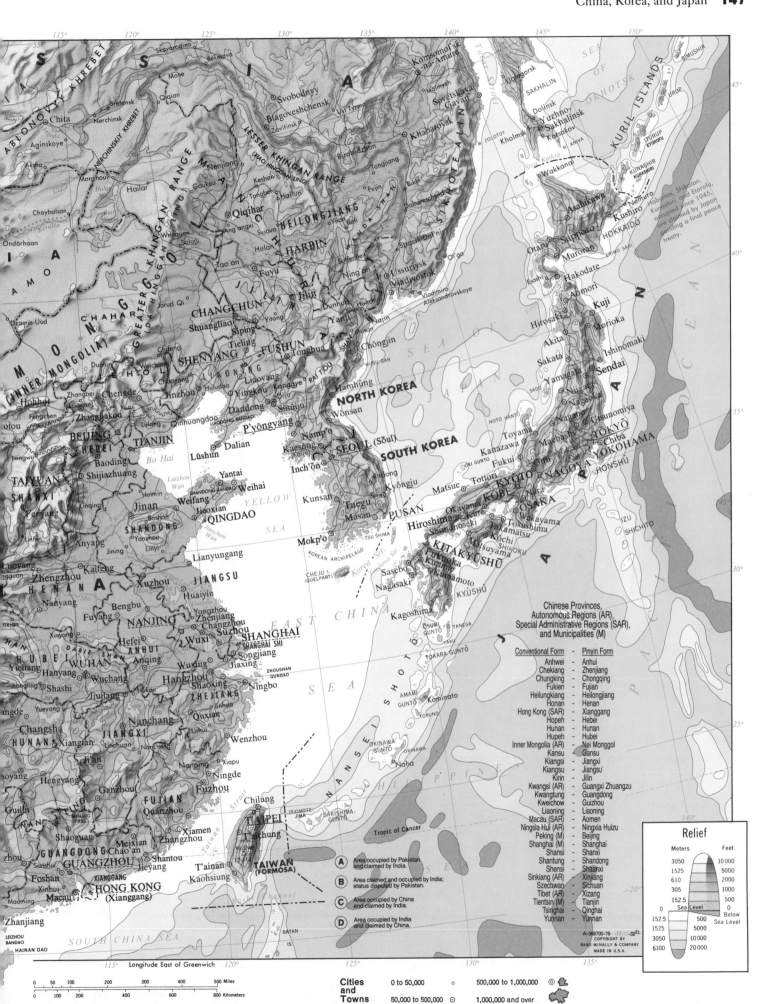

Chinese Provinces,
Autonomous Regions (AR),
Special Administrative Regions (SAR),
and Municipalities (M)

Conventional Form	Pinyin Form
Anhwei	Anhui
Chekiang	Zhenjiang
Chungking	Chongqing
Fukien	Fujian
Heilungkiang	Heilongjiang
Honan	Henan
Hong Kong (SAR)	Xianggang
Hopeh	Hebei
Hunan	Hunan
Hupeh	Hubei
Inner Mongolia (AR)	Nei Monggol
Kansu	Gansu
Kiangsi	Jiangxi
Kiangsu	Jiangsú
Kirin	Jilin
Kwangsi (AR)	Guangxi Zhuangzu
Kwangtung	Guangdong
Kweichow	Guizhou
Liaoning	Liaoning
Macau (SAR)	Aomen
Ningsia-Hui (AR)	Ningxia Huizu
Peking (M)	Beijing
Shanghai (M)	Shanghai
Shansi	Shanxi
Shantung	Shandong
Shensi	Shaanxi
Sinkiang (AR)	Xinjiang
Szechwan	Sichuan
Tibet (AR)	Xizang
Tientsin (M)	Tianjin
Tsinghai	Qinghai
Yunnan	Yunnan

Ⓐ Area occupied by Pakistan and claimed by India.

Ⓑ Area claimed and occupied by India; status disputed by Pakistan.

Ⓒ Area occupied by China and claimed by India.

Ⓓ Area occupied by India and claimed by China.

A-569700-76 -17-12-32EL
COPYRIGHT BY
RAND McNALLY & COMPANY
MADE IN U.S.A.

Relief

Meters		Feet
3050		10 000
1525		5000
610		2000
305		1000
152.5		500
0	Sea Level	0
		Below Sea Level
152.5		500
1525		5000
3050		10 000
6100		20 000

Longitude East of Greenwich

| 0 | 50 | 100 | 200 | 300 | 400 | 500 Miles |
| 0 | 100 | 200 | 400 | 600 | 800 Kilometers |

Cities and Towns

| 0 to 50,000 | ○ | 500,000 to 1,000,000 | ◎ |
| 50,000 to 500,000 | ⊙ | 1,000,000 and over | ▣ |

148

CHINA

Tropic of Cancer

MYANMAR (BURMA)
LAOS
THAILAND
CAMBODIA (KAMPUCHEA)
VIETNAM
MALAYSIA
BRUNEI
MALAYSIA
SUMATRA (SUMATERA)
BORNEO
KALIMANTAN
SINGAPORE
GREATER SUNDA ISLANDS
INDONE
JAVA (JAWA)
LESSER SUNDA ISLANDS

Relief

Meters		Feet
3050		10 000
1525		5000
610		2000
305		1000
152.5		500
	Sea Level	
152.5		500
1525		5000
3050		10 000
6100		20 000

A-569800-76 3-10-11-33EL
COPYRIGHT BY
RAND McNALLY & COMPANY
MADE IN U.S.A.

Longitude East of Greenwich

Scale 1:16 000 000; one inch to 250 miles. Polyconic Projection
Elevations and depressions are given in feet

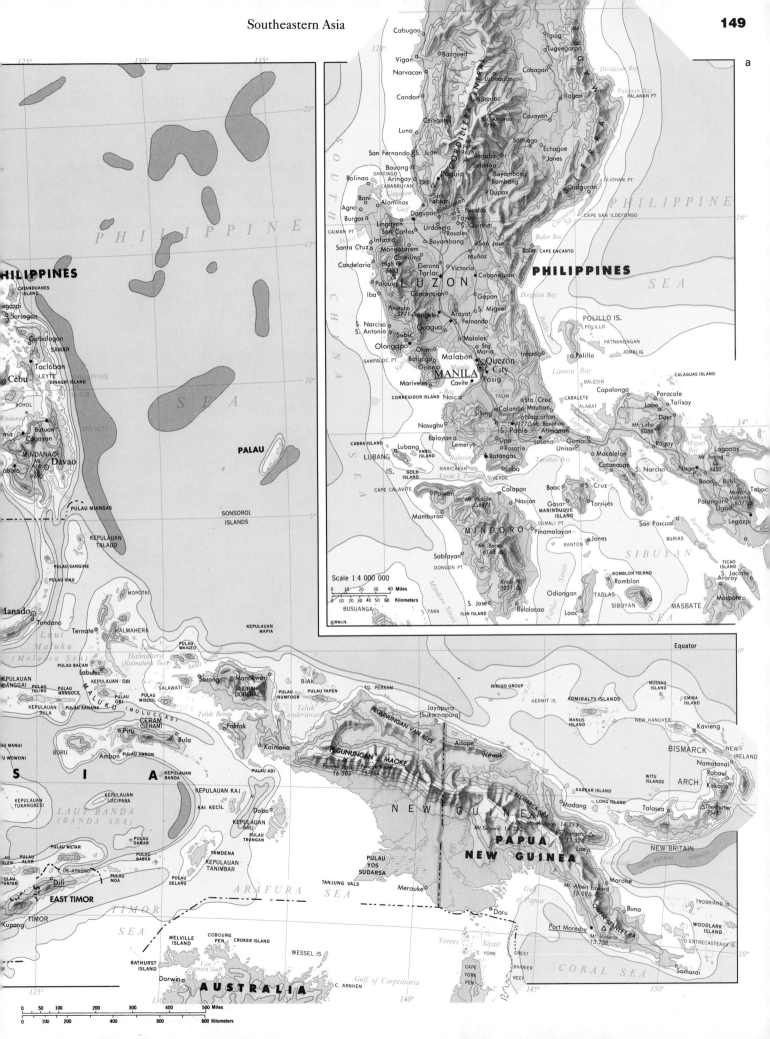

Oceania (including Australia and New Zealand)

Oceania is comprised of Australia, New Zealand, eastern New Guinea, and approximately 25,000 other islands in the South Pacific, most of which are uninhabited. Many of the islands are coral atolls, formed by microscopic creatures over scores of centuries, while others are the result of volcanic action.

Bay of Islands, North Island, New Zealand

Oceania's largest landmass is Australia, which at three million square miles (7.7 million sq km) is the world's smallest continent. In fact, it is smaller than five countries—Russia, Canada, China, Brazil, and the United States. Australia is generally flat and dry. The interior is sparsely populated, with most people living in coastal cities such as Sydney.

The next-largest part of Oceania is Papua New Guinea, the country occupying the eastern half of the island of New Guinea, which has some of the most forbidding and remote terrain in the world. New Zealand, Oceania's third-largest country, is known for its natural beauty and its huge herds of sheep.

Oceania at a glance

Land area: 3,300,000 square miles (8,500,000 sq km)

Estimated population: 31,415,000

Population density: 9.5/square mile (3.7/sq km)

Mean elevation: 1,000 feet (305 m)

Highest point: Mt. Wilhelm, Papua New Guinea, 14,793 feet (4,509 m)

Lowest point: Lake Eyre, South Australia, 52 feet (16 m) below sea level

Longest river: Murray-Darling, 2,330 mi (3,750 km)

Number of countries (incl. dependencies): 33

Largest independent country: Australia, 2,969,910 square miles (7,692,030 sq km)

Smallest independent country: Nauru, 8.0 square miles (21 sq km)

Most populous independent country: Australia, 19,455,000

Least populous independent country: Tuvalu, 11,000

Largest city: Brisbane, pop. 806,746

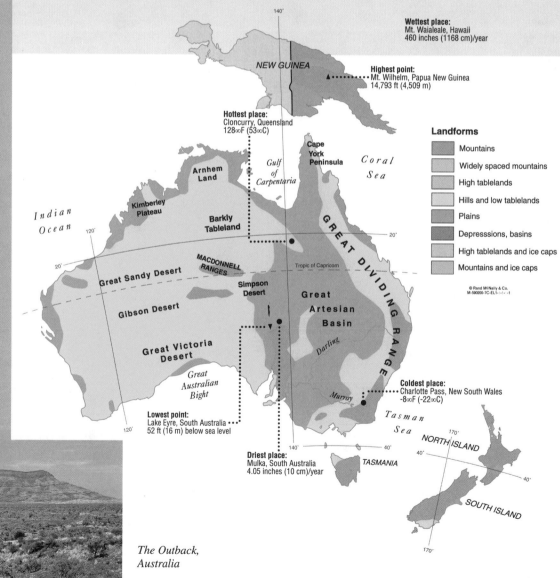

Wettest place:
Mt. Waialeale, Hawaii
460 inches (1168 cm)/year

NEW GUINEA

Highest point:
Mt. Wilhelm, Papua New Guinea
14,793 ft (4,509 m)

Hottest place:
Cloncurry, Queensland
128∞F (53∞C)

Cape York Peninsula

Coral Sea

Gulf of Carpentaria

Landforms

▢	Mountains
▢	Widely spaced mountains
▢	High tablelands
▢	Hills and low tablelands
▢	Plains
▢	Depresssions, basins
▢	High tablelands and ice caps
▢	Mountains and ice caps

© Rand McNally & Co.
M-590200-7C-EL1-·-·- -1

Arnhem Land

Indian Ocean

Kimberley Plateau

Barkly Tableland

GREAT DIVIDING RANGE

MACDONNELL RANGES

Tropic of Capricorn

Great Sandy Desert

Simpson Desert

Great Artesian Basin

Gibson Desert

Great Victoria Desert

Darling

Great Australian Bight

Coldest place:
Charlotte Pass, New South Wales
-8∞F (-22∞C)

Murray

Tasman Sea

Lowest point:
Lake Eyre, South Australia
52 ft (16 m) below sea level

NORTH ISLAND

TASMANIA

Driest place:
Mulka, South Australia
4.05 inches (10 cm)/year

SOUTH ISLAND

The Outback, Australia

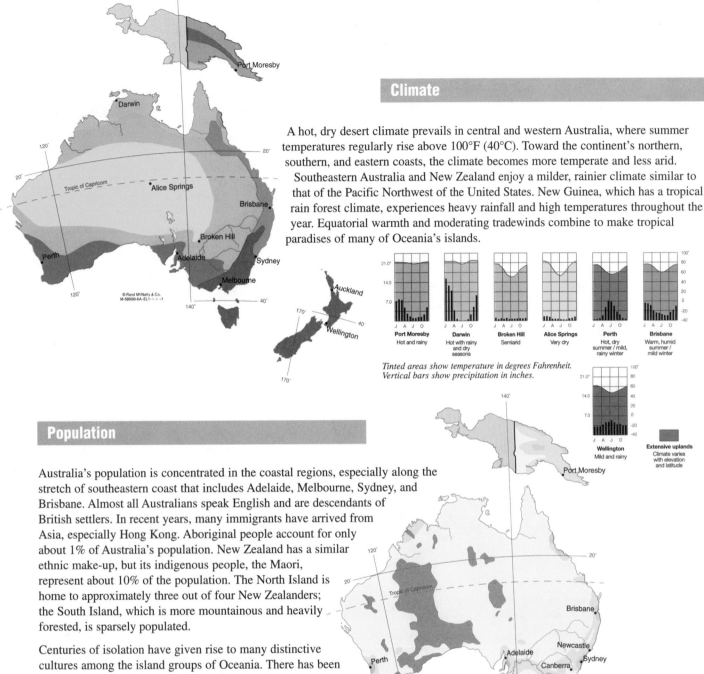

Climate

A hot, dry desert climate prevails in central and western Australia, where summer temperatures regularly rise above 100°F (40°C). Toward the continent's northern, southern, and eastern coasts, the climate becomes more temperate and less arid.

Southeastern Australia and New Zealand enjoy a milder, rainier climate similar to that of the Pacific Northwest of the United States. New Guinea, which has a tropical rain forest climate, experiences heavy rainfall and high temperatures throughout the year. Equatorial warmth and moderating tradewinds combine to make tropical paradises of many of Oceania's islands.

Port Moresby — Hot and rainy
Darwin — Hot with rainy and dry seasons
Broken Hill — Semiarid
Alice Springs — Very dry
Perth — Hot, dry summer / mild, rainy winter
Brisbane — Warm, humid summer / mild winter

Tinted areas show temperature in degrees Fahrenheit. Vertical bars show precipitation in inches.

Wellington — Mild and rainy

Extensive uplands — Climate varies with elevation and latitude

Population

Australia's population is concentrated in the coastal regions, especially along the stretch of southeastern coast that includes Adelaide, Melbourne, Sydney, and Brisbane. Almost all Australians speak English and are descendants of British settlers. In recent years, many immigrants have arrived from Asia, especially Hong Kong. Aboriginal people account for only about 1% of Australia's population. New Zealand has a similar ethnic make-up, but its indigenous people, the Maori, represent about 10% of the population. The North Island is home to approximately three out of four New Zealanders; the South Island, which is more mountainous and heavily forested, is sparsely populated.

Centuries of isolation have given rise to many distinctive cultures among the island groups of Oceania. There has been little immigration from other parts of the world.

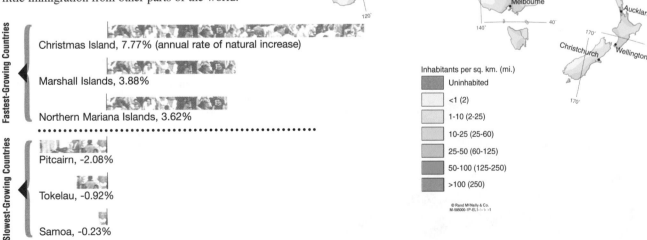

Fastest-Growing Countries

Christmas Island, 7.77% (annual rate of natural increase)

Marshall Islands, 3.88%

Northern Mariana Islands, 3.62%

Slowest-Growing Countries

Pitcairn, -2.08%

Tokelau, -0.92%

Samoa, -0.23%

Inhabitants per sq. km. (mi.)

- Uninhabited
- <1 (2)
- 1-10 (2-25)
- 10-25 (25-60)
- 25-50 (60-125)
- 50-100 (125-250)
- >100 (250)

© Rand McNally & Co.
M-595000-1P-EL1-

Environments and Land Use

Much of central and western Australia is a dry, inhospitable land of sand, rocks, and scrub vegetation. Surrounding this desert region is a broad band of semiarid grassland that covers more than half of the continent and supports a huge livestock industry. Australia has more sheep—132 million—than any other country in the world, as well as sizable herds of cattle. The dry climate and sparse plant life, however, mean that each animal requires a dozen or more acres to survive. Six percent of the continent is suitable for crops; most of the arable land is found on fertile plains in the southeast. Major crops include wheat, sugar cane, oats, barley, sorghum, and rice.

Farmland on North Island, New Zealand

Tourism plays an important role in Australia's economy. Among the continent's major attractions are its unusual wildlife, such as kangaroos, koalas, wombats, and platypuses; the Great Barrier Reef, which stretches for 1,250 miles (2,000 km) along the northeastern coast; and the ruggedly beautiful Outback, with its dramatic rock formations such as Ayers Rock (Uluru) and the Olga Rocks.

Thanks to its fertile land and temperate climate, New Zealand has a thriving livestock industry and is a leading world exporter of dairy products and lamb. Thinly populated and with little industry, it is one of the world's least polluted countries. Its pristine beauty encompasses a variety of scenery, including mountains, fjords, glaciers, rain forests, beaches and geysers. Only the country's relative isolation restrains its growing tourism industry.

Dense tropical rain forests blanket much of Papua New Guinea. These forests have thus far escaped the large-scale deforestation that is taking place in other tropical forests around the world.

Tourism is central to the economies of many of the islands throughout Oceania. Abundant sunshine, pleasant temperatures, and beautiful beaches draw millions of visitors each year to islands such as Tahiti and Fiji. For islands with little or no tourism, the economic scene is less promising: many islanders rely on subsistence fishing and foreign aid from former colonial powers.

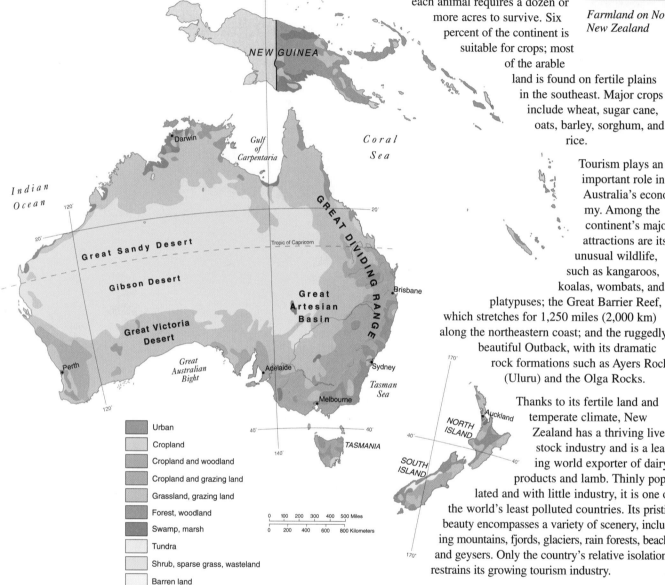

Urban
Cropland
Cropland and woodland
Cropland and grazing land
Grassland, grazing land
Forest, woodland
Swamp, marsh
Tundra
Shrub, sparse grass, wasteland
Barren land

© Rand McNally & Co.
N-ANS95000-M1- -2-2-1

Herding sheep near Goulburn, New South Wales

The Original Australians and New Zealanders

Many anthropologists believe that Australia's Aborigines are the oldest race of people on Earth. During the 40,000 years since they migrated to the island continent from Asia, they have developed a rich culture with an intricate spiritual and social life.

The original New Zealanders, the Maoris, arrived from other Polynesian islands in the 10th century. At the beginning of large-scale immigration from Britain in the 1800s, the British government signed a treaty with the Maoris which granted them full rights as citizens. With the exception of some disputes over land, this agreement has paved the way for the historically harmonious relations between the races in New Zealand.

Aboriginal boy with elders, Western Australia

Relations between the Aborigines and whites in Australia have been less harmonious. The arrival of the first European colonists in 1788 set in motion a chain of events that decimated the Aborigines and threatened their unique culture. Disease and skirmishes killed Aborigines along the coast, and thousands of others were forced from their lands by settlers. Some sought refuge with Aborigines already living in Australia's interior, the Outback. Alcoholism and other social problems became common among the Aborigines as they found themselves confronted by a society they did not understand.

Australia was slow to recognize the rights of its first inhabitants. In the early 1960s, official attitudes began to change as public embarrassment grew over the decades of discrimination. A significant step occurred in 1962 when full rights of citizenship were extended to the Aborigines.

Questions of land ownership, however, remain problematic for both

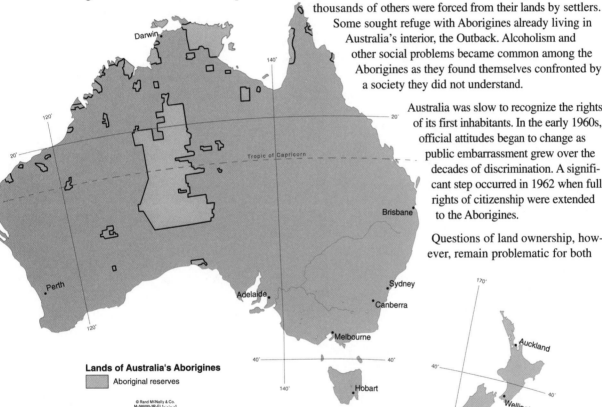

Lands of Australia's Aborigines

Aboriginal reserves

© Rand McNally & Co.
M-595000-3R-EL1-·-·-·- -1

sides. The government has set aside large reserves for the Aborigines, but much of the land is in the continent's hostile interior (see map above). Only in the past two decades has an agreement been reached allowing Aborigines to share in the vast mineral wealth of their northern lands. Recent court decisions have awarded individual Aborigines rights to ancestral lands which were seized by settlers, but many local governments continue to fight these decisions.

In the face of indifference and hostility, there has recently been an upsurge of interest in cultural traditions among the approximately 240,000 Australians of aboriginal descent. Still, many of the traditions of the 500 different tribes that were present 200 years ago have been lost.

INDONESIA

JAVA Pasuruan 10 932
G. Mahameru 12 060 G. Raung
Singaraja Rinjani 8225
Bali Selat Sumbawa Raba
LOMBOK Besar Alor Dili EAST TIMOR
SUMBAWA FLORES LOMBLEN PANTAR
Waingapu SAVU
SUMBA SEA TIMOR
SAWU ROTI Kupang

SELARU TANJUNG VALS

A R A F U R A   S E A

S U N D A   I S L A N D S

T I M O R   S E A

SUNDA TRENCH

SELARU

C. VAN DIEMEN CROKER WESSEL IS.
BATHURST MELVILLE COBURG PEN.
Van Diemen Gulf CAPE ARNHEM
Darwin Clarence Str.
Pine Creek ARNHEM LAND Blue Mud Bay
GROOTE EYLANDT
CAPE Katherine Roper Limmen
LONDONDERRY Joseph Bight
Bonaparte Gulf SIR EDWARD PELLEW GROUP
WELLESLEY
Birdum Borroloola
Victoria River Daly Waters
Downs Newcastle Waters Burketown
NORTHERN Alexandria BARKLY TABLELAND
Tanami Tennant Creek Dobbyn
Camooweal
T E R R I T O R Y Mount Isa
Malbd
Barrow Creek Dajarra
QU
Mt. Ziel Arltunga
4955 RANGES
MACDONNELL Alice Springs
JAMES RANGE SIMPSON
Uluru Amadeus DESERT
(Ayers Rock) Charlotte Birdsville
Waters A
MUSGRAVE RANGES
Mt. Woodroffe EVERARD RANGES
4724 Oodnadatta
William Creek
STUART RANGE Marree
SOUTH AUSTRALIA
Farind
Woomera
Parachilna
Pimba FLINDERS RANGES
Ooldea Station FLIND
Hughes Penong Port Augusta
Eucla Ceduna Whyalla Peterbor
POINT FOWLER EYRE Port Pirie Gladstone
PENINSULA Moonta Wallaroo
Port Wakefield
Port Lincoln Gawler
Adela
KANGAROO Murr
Bri

BUCCANEER ARCH.
CAPE LEVEQUE
DAMPIER Derby
LAND GEIKIE KING LEOPOLD RANGES
Broome RANGE Mt. Hann
Roebuck Bay Fitzroy 2800
LaGrange Crossing Halls Creek
Wyndham
EIGHTY MILE BEACH
LARREY POINT

DAMPIER RIPON Port Hedland
ARCH. De Grey
MONTE BELLO IS. Roebourne
BARROW Marble Bar
NORTH WEST CAPE Nullagine
POINT CLOATES Millstream Onslow
HAMERSLEY RANGE Jiggalong
Mt. Bruce 4052
TROPIC OF CAPRICORN
CAPE FARQUHAR Carnarvon
BERNIER Peak Hill
DORRE Nabberu Carnegie Gillen
Gascoyne Wells
DIRK HARTOG Meekatharra Wiluna
STEEP POINT Cue Sandstone Yeo
Ajana Mount Laverton
Northampton Magnet Rawlinna
HOUTMAN ROCKS Menzies Hughes
Geraldton GREAT VICTORIA DESERT
Dongara Mingenew Kalgoorlie-Boulder Eyre
Pithara Coolgardie
DARLING Miling Lake Brown Southern Cross
Moora York Norseman Salmon Gums
Perth Northam Dundas
Fremantle Narrogin Ravensthorpe Esperance
Collie Hopetoun ARCHIPELAGO
Geographe Bay OF THE RECHERCHE
Bunbury
CAPE NATURALISTE Busselton Katanning
CAPE LEEUWIN Nornalup Albany
PT. D'ENTRECASTEAUX West Cape Howe King George Sd.

GREAT SANDY DESERT Mackay
Disappointment
GIBSON DESERT Macdonnell
WESTERN
AUSTRALIA
NULLARBOR PLAIN
GREAT AUSTRALIAN BIGHT
SWANLAND

I N D I A N   O C E A N

CAPE JAFFA Naraco
Kingston
Mt. Gam

40,000 SQ MI
AREA

0   100   200
Miles

A-590200-26 5-18EL
COPYRIGHT BY
RAND McNALLY & COMPANY
MADE IN U.S.A.

Longitude 115° East of Greenwich

Scale 1:16 000 000; one inch to 250 miles. Lambert's Azimuthal, Equal Area Projection
Elevations and depressions are given in feet

155°

CHOISEUL

VELLA
LAVELLA

160°   SANTA ISABEL

165°

170°

EW GUINEA
PAPUA NEW GUINEA
Mt. Albert Edward
13,100
Buna
Port Moresby
Mt. Victoria
13,363
OWEN STANLEY RA.

TROBRIAND IS.
WOODLARK
D'ENTRECASTEAUX
ISLANDS

RENDOVA
NEW
GEORGIA

MALAITA
Honiara
SOLOMON   ISLANDS
FLORIDA
TULAGI
RUSSELL IS.

Torres   Strait
ULGRAVE
BANKS
THURSDAY
HORN
RINCE OF
WALES
CAPE YORK

SOUTH CAPE
Samarai
LOUISIADE
ARCHIPELAGO

GUADALCANAL

SAN CRISTÓBAL

SANTA CRUZ
ISLANDS

eipa
CAPE
YORK
PENINSULA

TAGULA
ROSSEL

RENNELL

10°

TORRES IS.

BANKS
ISLANDS

Laura
Cooktown
Palmerville
ATHERTON
Mungand
Croydon
Forsayth
PLATEAU
Inghom
HINCHINBROOK
Townsville
Halifax Bay
Charters
Towers
Bowen

ormanton
Richmond
Hughenden
Kynuna

Cairns
Mt. Barfle Frere
5322

GREAT

OSPREY REEF

CAPE MELVILLE

HOLMES
REEFS

FLINDERS
REEFS

WILLIS IS.

TREGROSSE IS.

MARION REEF

C O R A L   S E A

ESPÍRITU SANTO

NEW

MALEKULA

HEBRIDES

MAEWO

PENTECOST

AMBRIM
EPI
EEFATE
Port Vila

AMBRIM

VANUATU

15°

Whitsunday
CUMBERLAND IS.
Mackay

BARRIER   REEF

Mt. Dalrymple
4190
NORTHUMBERLAND IS.
SWAIN REEFS

Í LES CHESTERFIELD
(Fr.)

ÍLES BÉLEP

P A C I F I C

TANA

EROMANGA

ANEITYUM

20°

ENSLAND
GREAT
Barcaldine
Longreach
Clermont
Emerald
Jericho
Ding
Yaraka
Blackall
Tambo
BUCKLAND
TABLELAND

TESIANGE

Windorah
Quilpie

CONNORS RANGE

Rockhampton
Mount Morgan
CURTIS
Gladstone

Capricorn Chan.

WRECK REEFS

OUVÉA
LIFOU
ÍLES LOYAUTÉ
MARÉ
(French)

**NEW
CALEDONIA**
(Fr.)

Nouméa

ÍLE DES PINS

Tropic of Capricorn

amma Yamma

Bundaberg
Hervey
Bay
Maryborough
FRASER
Fraser

SANDY CAPE

O C E A N

25°

Thargomindah
St. George
Charleville
Roma

Gympie

160°

165°

170°

Cunnamulla

GREEN

DARLING
DOWNS
Toowoomba
Dalby
Ipswich
Brisbane
Warwick
Southport
N. STRADBROKE I.

a

Hungerford
Dirranbandi
Mungindi
Mt. Roberts
4495
Lismore

34°

Brewarrina
Bourke
Moree
Narrabri
Inverell
Mt. Roberts
5100
NEW
ENGLAND
RANGE
Grafton

PACIFIC

NORTH CAPE
Kaitaia
Russell

Walgett

Tenterfield

NORTH ISLAND

GREAT
BARRIER

**MAIN
BARRIER
RANGE**

Cobar
Coonamble
Armidale
Tamworth
5360
The Round Mountain

WARRUMBUNGLE
RA.

Kempsey

OCEAN

Devonport
Auckland

Bay of
Plenty

Hamilton
EAST CAPE

Broken Hill
Wilcannia
Nyngan
Dubbo

LIVERPOOL
RA.

Port Macquarie

38°

NEW SOUTH WALES
Nymagee

Forbes
Cessnock
Maitland
Newcastle

North Taranaki Bight
New Plymouth
C. EGMONT
Mt. Egmont
Vol.

Hawke Bay

Gisborne

nmark
MURRAY
Wentworth
Hay
West
Wyalong
Bathurst
Orange
Lithgow
BLUE MTS.
SYDNEY
Botany Bay

**NEW
ZEALAND**

Wanganui

Napier
Hastings
Palmerston North

RIVERINA

Narrandera
Goulburn
Jervis Bay

Wollongong

South Taranaki Bight

ebinga
ildura

REGION

Wagga
Wagga

Albury

AUSTL. CAP. TER.
Canberra
Cooma

CAPE
FAREWELL
Tasman
Bay
Nelson

Cook

Lower Hutt
Wellington

35°

Swan Hill
Kerang
Deniliquin

Echuca

MURRAY

Mt. Kosciusko
7316
SNOWY
MTS.

Bega
Bombala

T A S M A N

**T A S M A N
S E A**

Karamea Bight

CAPE FOULWIND

Strait

42°

orsham
VICTORIA

Bendigo

Benalla

GREAT

Greymouth
Hokitika

Pegasus Bay

Ararat
Maryborough
SNOWY
MTS.

Ballarat

CAPE HOWE

SOUTH ISLAND

SOUTHERN ALPS

Christchurch

amilton
Geelong
MELBOURNE
Bairnsdale
NINETY MILE BEACH

S E A

ortland
Warrnambool
Wonthaggi

CAPE OTWAY
WILSON'S
PROMONTORY

CASCADE PT.

Canterbury Bight

PACIFIC

KING
FLINDERS

Timaru

RESOLUTION
ISLAND

Foveaux

Dunedin
CAPE SAUNDERS

46°

TASMANIA
Burnie
Ulverstone
Devonport
HUNTER IS.
FURNEAUX GROUP
CAPE BARREN

STEWART ISLAND

Invercargill

SOUTHWEST
CAPE

O C E A N

Launceston
Mt. Ossa
5305
Strahan

New Norfolk
Risdon
Hobart
BRUNY
SOUTH EAST CAPE

®RMCN.

Same scale as main map

168°

172°

176°

50   100   200   300   400   500 Miles
100   200   400   600   800 Kilometers

**Cities
and
Towns**

0 to 50,000

50,000 to 500,000

500,000 to 1,000,000

1,000,000 and over

156

Relief

Meters	Feet
3050	10 000
1525	5 000
610	2000
305	1000
152.5	500
0 Sea Level	0
152.5	500 Below Sea Level
1525	5000
3050	10 000
6100	20 000

A-590200-76   7   -18^{EL}
COPYRIGHT BY
RAND McNALLY & COMPANY
MADE IN U.S.A.

Scale 1:16 000 000; one inch to 250 miles. Lambert's Azimuthal, Equal Area Projection
Elevations and depressions are given in feet

Longitude   East of Greenwich

INDONESIA

Pasuruan
G. Mahameru 12 060
G. Raoeng 10 932
Singaraja
Bali
Lombok
SUMBAWA
Sumbawa Besar
Bima
FLORES
Waingapu
SUMBA
SAWU
ROTI
Kupang
TIMOR
Dili
EAST TIMOR
ALOR
LOMBLEN PANTAR
SELARU
TANJUNG VALS

SUNDA   ISLANDS

SAVU SEA

ARAFURA   SEA

TIMOR   SEA

SUNDA   TRENCH

INDIAN   OCEAN

CAPE LONDONDERRY
CAPE VAN DIEMEN
BATHURST
MELVILLE
Van Diemen Gulf
Clarence Str.
Darwin
CROKER
COBURG PEN
WESSEL IS.
CAPE ARNHEM
Anson Bay
Blue Mud Bay
GROOTE EYLANDT
Limmen Bight
GULF OF CARPENTARIA

Joseph Bonaparte Gulf
Queens Chan.
ARNHEM LAND
Pine Creek
Katherine

SIR EDWARD PELLEW GROUP
WELLESLEY IS.

Wyndham
Mt. Hann 2800
KING LEOPOLD RANGES
GEIKIE RANGE
Fitzroy Crossing
Halls Creek
Derby
BROOME LAND
DAMPIER LAND
CAPE LEVEQUE
BUCCANEER
King Sound
ARCH.
Collier Bay
Sunday ARCH.
Roebuck Bay
LaGrange
Broome

Birdum
Victoria River Downs
Daly Waters
Newcastle Waters
Borroloola
Burketown
BARKLY TABLELAND

NORTHERN

TERRITORY

QU

Tanami
Tennant Creek
Alexandria
Dobbyn
Camooweal
Mount Isa
Malbo
Duc
Dajarra

Barrow Creek

EIGHTY MILE BEACH
LARREY POINT
RIPON
Port Hedland
DAMPIER ARCH.
DeGrey
Roebourne
Marble Bar
Nullagine
GREAT SANDY DESERT
Mackay

MONTE BELLO IS.
BARROW
NORTH WEST CAPE
Onslow
Millstream
HAMERSLEY RANGE
Mt. Bruce 4052
Jiggalong
Mt. Ziel 4955
MACDONNELL RANGES
Arltunga
Alice Springs
JAMES RANGE
SIMPSON DESERT
Birdsville

POINT CLOATES
Tropic of Capricorn
CAPE FARQUHAR
CAPE Geographe
Carnarvon
Gascoyne
Peak Hill
Nabberu
Carnegie
GIBSON DESERT
Gillen
Uluru (Ayers Rock)
Charlotte Waters
MUSGRAVE RANGES
Mt. Woodroffe 4724
EVERARD RANGES
Oodnadatta

BERNIER
DORRE
Shark Bay
DIRK HARTOG
STEEP POINT
Meekatharra
Nannine
Cue
Sandstone
Austin
Wiluna
Welld
Gillen
STUART RANGE
William Creek

WESTERN

AUSTRALIA

Ajana
Mount Magnet
Laverton
GREAT VICTORIA DESERT
SOUTH   AUSTRALIA
Marree
Parina

Northampton
Ballard
Menzies
Area
Oldea Station
Woomera
Parachilna
Pimba

HOUTMAN ROCKS
Geraldton
Mingenew
Moore
Barlee
Kalgoorlie-Boulder
Coolgardie
Rawlinna
Hughes
NULLARBOR   PLAIN
Penong
Ceduna
Everard
FLIND
Port Augusta
Whyalla
Port Pirie
Peterbor
Gladstone

Dongara
DARLING RANGE
Pithara
Miling
Mooral
Lake Brown
SWANLAND
Southern Cross
Cowan
Norseman
Dundas
Eucla
Eyre
POINT FOWLER
EYRE PENINSULA
Moonta
Port Wakefie
Wallaroo
Gawler
Adela
Murr
Brid

Perth
Fremantle
Northam
York
Narrogin
Collie
Bunbury
CAPE NATURALISTE
Busselton
Katanning
Ravensthorpe
Esperance
Salmon Gums
Geographe Bay
Hopetoun
ARCHIPELAGO OF THE RECHERCHE
GREAT   AUSTRALIAN   BIGHT
Port Lincoln
KANGAROO
Naraco
Kingston
CAPE JAFFA
Mt. Gamb

CAPE LEEUWIN
Nornalup
Albany
PT. D'ENTRECASTEAUX
WEST CAPE HOWE
King George Sd.

INDIAN   OCEAN

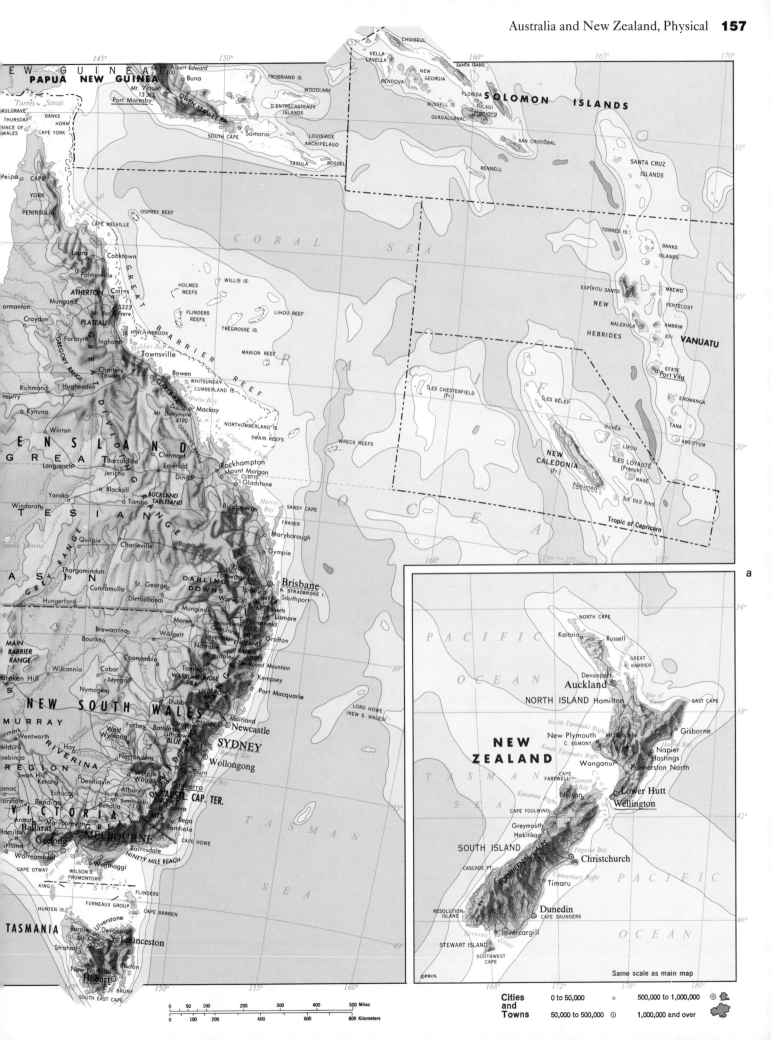

RUSSIA

KAZAKHSTAN
Irkutsk
ZAPADNYYE SAYAN
Lake Baikal
STANOVOY KHREBET
SEA OF OKHOTSK
KOMANDORSKIYE OSTROVA
Petropavlovsk-Kamchatskiy
Nome
ST. LAWRENCE
ALA (U.S.A.)
Unalaska
ALEUTIAN IS.
ALASKA

Ulan Bator
MONGOLIA
GOBI DESERT
MANCHURIA
GREATER KHINGAN RANGE
HARBIN
CHANGCHUN
SHENYANG
MS. LOPATKA
KURILE IS.
HOKKAIDO
Vladivostok

BEIJING
TIANJIN
Dalian
KOREA
SEOUL (Sŏul)
HONSHU
TŌKYŌ
SEA OF JAPAN
JAPAN CURRENT

KUNLUN SHAN
C H I N A
KŌBE
Nagasaki
YOKOHAMA
KITAKYŪSHŪ
KYŪSHŪ
NANJING
WUHAN
SHANGHAI
Yangtze

Fuzhou
T'AIPEI
NANSEI SHOTO
TAIWAN (FORMOSA)
BONIN IS. (Japan)
Tropic of Cancer
MARCUS (Japan)
MIDWAY IS. (U.S.A.)
INTERNATIONAL DATE LINE

GUANGZHOU
HONG KONG
Hanoi
CAPE ENGANO
Hue
HAINAN DAO
PHILIPPINE SEA
MARIANA IS.
NORTHERN MARIANA ISLANDS (U.S.A.)
WAKE (U.S.A.)

THAILAND
BANGKOK
LAOS
VIETNAM
SOUTH CHINA SEA
MANILA
PHILIPPINES
LUZON
GUAM (U.S.A.)
NORTH EQUATORIAL CURRENT

CAMBODIA
Gulf of Thailand
SAMAR
MARSHALL IS.
MARSHALL ISLANDS

HO CHI MINH CITY (Saigon)
MINDANAO
CAROLINE IS.
FEDERATED STATES OF MICRONESIA

MALAY PENINSULA
Bandar Seri Begawan
BRUNEI
PALAU IS.
M I C R O N E S I A

MALAYSIA
MALAYSIA
CELEBES SEA
PALAU
HALMAHERA
Equator
NAURU
GILBERT IS.
HOWLAND (U.S.A.)
BAKER (U.S.A.)
KIRIBATI

SINGAPORE
SINGAPORE
SUMATRA
BORNEO
CELEBES
MOLUCCAS
Manokwari
TG. PERRAM
Jayapura (Sukarnapura)
NEW IRELAND
BISMARCK ARCH.
KANTON
PHOENIX IS.
ENDERBU

INDONESIA
JAKARTA
JAVA SEA
CERAM
PAPUA NEW GUINEA
NEW BRITAIN
M E L A N E S I A
TUVALU
SOLOMON ISLANDS
BOUGAINVILLE
TOKELAU (N.Z.)

JAVA TRENCH
ARAFURA SEA
TIMOR
EAST TIMOR
THURSDAY
CAPE YORK
Port Moresby
SOUTH CAPE
CORAL SEA
WALLIS AND FUTUNA (Fr.)
SAMOA
AMER. SAMOA

CHRISTMAS (Austl.)
TIMOR SEA
Darwin
Gulf of Carpentaria
NEW HEBRIDES
VANUATU
FIJI
TONGA

NORTH WEST CAPE
GREAT SANDY DESERT
Tropic of Capricorn
MACDONNELL RANGES
A U S T R A L I A
GREAT DIVIDING RANGE
EAST AUSTRALIAN CURRENT
LOYALTY IS.
NEW CALEDONIA (Fr.)

Perth
Fremantle
Albany
Great Australian Bight
Torrens
Brisbane
NORFOLK (Austl.)
KERMADEC IS. (N.Z.)
NORTH CAPE
NORTH ISLAND

Adelaide
Canberra
SYDNEY
Murray
TASMAN SEA
Auckland

MELBOURNE
CAPE HOWE
Bass Strait
TASMANIA
Hobart
SOUTH EAST CAPE
SOUTH ISLAND
NEW ZEALAND
Wellington

INDIAN OCEAN
Dunedin
STEWART
SOUTHWEST CAPE
CHAT IS. (N.Z.)

BEIRING SEA
ATTU

Relief

Meters		Feet
3050		10 000
1525		5000
610		2000
305		1000
152.5		500
0	Sea Level	0
152.5		500
1525		5000
3050		10 000
6100		20 000

A-598500-76
COPYRIGHT BY
RAND McNALLY & COMPANY
MADE IN U.S.A.

70° 80° 90° 100° 110° 120° Longitude 130° East of 140° Greenwich 150° 160° 170°

→ Warm ocean currents
→ Cold ocean currents

Scale 1:50 000 000; one inch to 800 miles. Goode's Homolosine Equal Area Projection
Elevations and depressions are given in feet

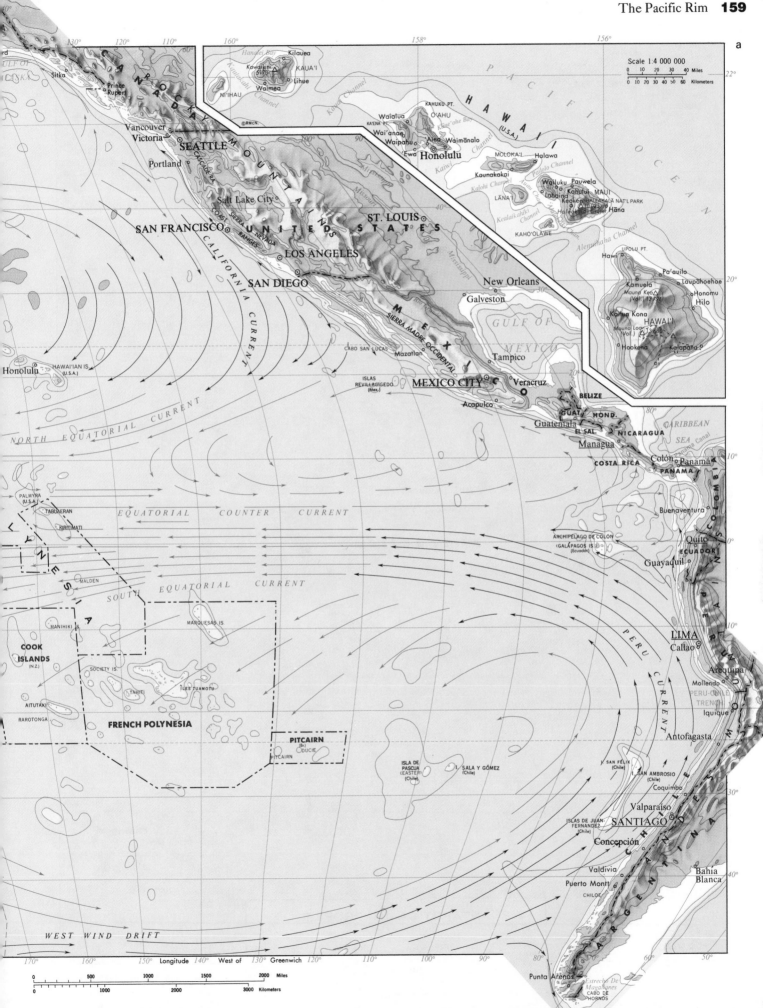

a

Scale 1:4 000 000

0 10 20 30 40 Miles
0 10 20 30 40 50 60 Kilometers

HAWAI'I
(U.S.A.)

Hanalei Bay   Kilauea
Kawaikini △   **KAUA'I**
(5170)
NI'IHAU   Lihue
Waimea

©RMCN.

KAHUKU PT.
Walalua   **O'AHU**
KA'ENA PT.   Kāne'ohe Bay
Wai'anae   Aiea   Waimānalo
Waipahu   Kailua
Ewa   **Honolulu**

MOLOKA'I   Halawa
Kaunakakai
Kalohi Channel   Wailuku   Pauwela
Pailolo Channel   Kahului   **MAUI**
LĀNA'I   Lahaina   Keakea   HALEAKALA NAT'L PARK
Haldea   Pa'ahana   Hāna
Kealaikahiki Channel   HALEAKALA △
Channel
KAHO'OLAWE   Alenuihaha Channel

UPOLU PT.
Hawi   Pa'auilo
Kamuela   Laupahoehoe
Mauna Kea △ 13,796   Honomu
Kailua Kona   Hilo
Mauna Loa   Kalapana
(Vol.) △ 13,680   Hookena   Kalapana

PACIFIC

CANADA

GULF OF
ard   ALASKA
Sitka
Prince
Rupert
Vancouver
Victoria
SEATTLE
Portland   CASCADE RA.
ROCKY MOUNTAINS
Salt Lake City
COAST
SAN FRANCISCO   RANGES
SIERRA NEVADA   **UNITED STATES**   **ST. LOUIS**
LOS ANGELES
CALIFORNIA CURRENT   **SAN DIEGO**
New Orleans
Galveston
GULF OF
CABO SAN LUCAS
MEXICO
Mazatlan
Tampico
ISLAS REVILLAGIGEDO   SIERRA MADRE OCCIDENTAL   **MEXICO CITY**   Veracruz
(Mex.)
BELIZE
Acapulco   **GUAT.**   **HOND.**
Honolulu   HAWAI'IAN IS.   **Guatemala**   CARIBBEAN
(U.S.A.)   **EL SAL.**   **NICARAGUA**   SEA
Managua
NORTH   EQUATORIAL   CURRENT   **COSTA RICA**   Colón   **Panamá**
PANAMA
Buenaventura
PALMYRA   ARCHIPÉLAGO DE COLÓN   Quito
(U.S.A.)   (GALÁPAGOS IS.)   **ECUADOR**
TABUAERAN   EQUATORIAL   COUNTER   CURRENT   (Ecuador)   Guayaquil
KIRITIMATI
COLOMBIA
SOUTH   EQUATORIAL   CURRENT
MALDEN   MARQUESAS IS.   **LIMA**
MANIHIKI IS.   Callao
COOK   PERU
ISLANDS   SOCIETY IS.   Arequipa
(N.Z.)   TAHITI   ÎLES TUAMOTU   Mollendo
PERU-CHILE
AITUTAKI   TRENCH
RAROTONGA   Iquique
FRENCH POLYNESIA
PITCAIRN   Antofagasta
(Br.)
DUCIE
PITCAIRN   I. SAN FÉLIX
(Chile)   I. SAN AMBROSIO
ISLA DE   (Chile)
PASCUA   I. SALA Y GÓMEZ   Coquimbo
(EASTER)   (Chile)
(Chile)   ANDES   CHILE
Valparaíso
ISLAS DE JUAN   **SANTIAGO**
FERNANDEZ   ARGENTINA
(Chile)
Concepción
Valdivia   Bahia
Puerto Montt   Blanca
CHILOÉ
WEST   WIND   DRIFT
Punta Arenas   Estrecho De
Magallanes
CABO DE
HORNOS

POLYNESIA

PERU CURRENT

170° 160° 150° Longitude 140° West of 130° Greenwich 120° 110° 100° 90° 80° 70° 60° 50°

0   500   1000   1500   2000 Miles
0   1000   2000   3000 Kilometers

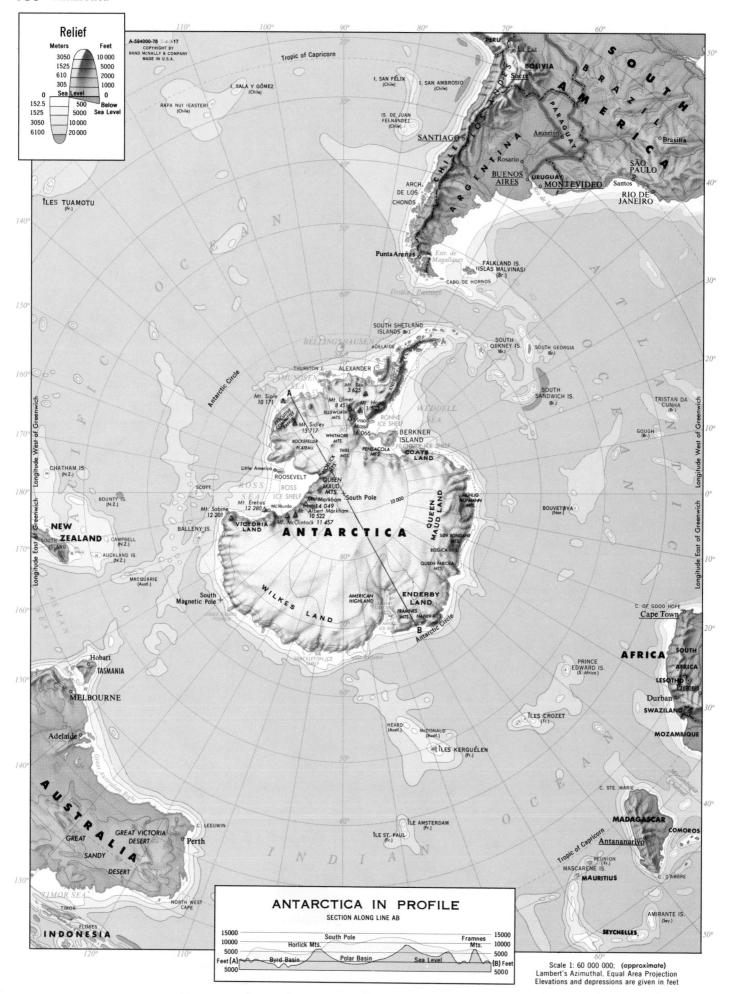

Glossary

Foreign Geographical Terms

Afk. ... Afrikaans
Ara. ... Arabic
Ber. ... Berber
Blg. ... Bulgarian
Bur. ... Burmese
Cbd. ... Cambodian
Ch. ... Chinese
Czech ... Czech
Dan. ... Danish
Du. ... Dutch
Est. ... Estonian
Finn. ... Finnish
Fr. ... French
Gae. ... Gaelic
Ger. ... German
Gr. ... Greek
Heb. ... Hebrew
Ice. ... Icelandic
Indon. ... Indonesian
It. ... Italian
Jpn. ... Japanese
Kor. ... Korean
Lao. ... Laotian
Lapp. ... Lappish
Mal. ... Malay
Mong. ... Mongolian
Nor. ... Norwegian
Pas. ... Pashto
Per. ... Persian
Pol. ... Polish
Port. ... Portuguese
Rom. ... Romanian
Rus. ... Russian
S./C. ... Serbo-Croatian
Slo. ... Slovak
Sp. ... Spanish
Swe. ... Swedish
Thai ... Thai
Tib. ... Tibetan
Tur. ... Turkish
Ukr. ... Ukranian
Viet. ... Vietnamese

-å, Dan., Nor., Swe. ... river
āb, Per. ... river
ada(lar), Tur. ... island(s)
adrar, Ber. ... mountains
ákra, akrotírion, Gr. ... cape
altos, Sp. ... mountains, hills
-älv,-älven, Swe. ... river
-ån, Swe. ... river
archipel, Fr. ... archipelago
archipiélago, Sp. ... archipelago
arquipélago, Port. ... archipelago
arroyo, Sp. ... brook
-ås,-åsen, Swe. ... hills
baai, Du. ... bay
bab, Ara. ... strait
Bach, Ger. ... brook, creek
-backen, Swe. ... hill
bælt, Dan. ... strait
bahía, Sp. ... bay
bahr, baḩr, Ara. ... river, sea
baía, Port. ... bay
baie, Fr. ... bay
-bana, Jpn. ... cape
banco, Sp. ... bank
bandao, Ch. ... peninsula
bassin, Fr. ... basin
batang, Indon. ... river
bātlāq, Per. ... marsh
ben, Gae. ... mountain
Berg, Ger. ... mountain, hill
-berg, Afk. ... mountains
Berge, Ger. ... mountains
bi'r, Ara. ... well
birkat, Ara. ... lake
bocca, It. ... river mouth, pass
boğazı, Tur. ... strait
bogd, Mong. ... range
bolsón, Sp. ... enclosed basin
-breen, Nor. ... glacier
Brücke, Ger. ... bridge
Bucht, Ger. ... bay
bugt, Dan. ... bay
bukit, Indon., Mal. .. mountain, hill
-bukten, Swe. ... bay
bulu, Indon. ... mountain
Burg, Ger. ... castle
burn, Gae. ... brook
burnu, burun, Tur. ... cape
cabezas, Sp. ... peaks
cabo, Port., Sp. ... cape
campo, It. ... plain
cap, Fr. ... cape
capo, It. ... cape
catena, Sp. ... range
cayo(s), Sp. ... cay(s), islet(s)
cerro(s), Sp. ... mountain(s), hill(s)
chaîne, Fr. ... range
château, Fr. ... castle
chiang, Ch. ... harbor, harbour
chott, Ara. ... intermittent lake, salt marsh
cima, It., Sp. ... peak
città, It. ... city
ciudad, Sp. ... city
co, Tib. ... lake
co., cerro, Sp. ... mountain, hill
col, Fr. ... pass
colina(s), Sp. ... hill(s)
colline, It. ... hills
collines, Fr. ... hills
con, Viet. ... islands
cord., cordillera, Sp. ... range
costa, Sp. ... coast
côte, Fr. ... coast, hills
cuchilla, Sp. ... hills, ridge
dağ, dağı, Tur. ... mountain
dāgh, Per. ... mountains
-dake, Jpn. ... mountain
-dal, -dalen, Nor., Swe. ... valley
danau, Indon. ... lake
dao, Ch., Viet. ... island
daryācheh, Per. ... lake
dasht, Per. ... desert
deniz, denizi, Tur. ... sea
desierto, Sp. ... desert
détroit, Fr. ... strait
dijk, Du. ... dike
distrito, Sp. ... district
djebel, Ara. ... mountain(s)
-do, Kor. ... island
-elv,-elva, Nor. ... river
embalse, Sp. ... reservoir
erg, Ara. ... sand desert
estrecho, Sp. ... strait
étang, Fr. ... pond
-ey, Ice. ... island
fjäll(en), Swe. ... mountain(s)
fjället, Swe. ... mountain
fjärden, Swe. ... fjord
-fjell, -fjellet, Nor. ... mountain
-fjord, Nor. ... fjord
-fjorden, Nor., Swe. ... fjord, lake
-fjörur, Ice. ... fjord, bay
-flói, Ice. ... bay
foce, It. ... river mouth, pass
forêt, Fr. ... forest
-forsen, Swe. ... waterfall
Forst, Ger. ... forest
-foss, Ice. ... waterfall
-fossen, Nor. ... waterfall
g., gora, Rus. ... mountain, hill
g., gunong, Mal. ... mountain
gang, Ch. ... bay
-gang, Kor. ... river
gave, Fr. ... mountain torrent
gebergte, Du. ... range
Gebirge, Ger. ... range
Gipfel, Ger. ... peak
göl, Tur. ... lake
golfe, Fr. ... gulf
golfete, Sp. ... bay
golfo, It., Sp. ... gulf
gölü, Tur. ... lake
gora, Rus. ... mountain, hill
gora, S. \C. ... mountains
góra, Pol. ... mountain
gory, Rus. ... mountains, hills
góry, Pol. ... mountains
gr'ada, Rus. ... ridge
guba, Rus. ... bay
gunong, Mal. ... mountain
gunung, Indon. ... mountain
-guntō, Jpn. ... islands
Haff, Ger. ... lagoon
hai, Ch. ... sea, lake
-hama, Jpn. ... beach
hamada, Ara. ... desert
hāmūn, Per. ... lake, marsh
-hantō, Jpn. ... peninsula
hare, Heb. ... mountains, hills
-hav, Swe. ... sea, bay
havre, Fr. ... harbor, harbour
he, Ch. ... river
ho, Ch. ... river
-ho, Kor. ... reservoir
-holm, Dan. ... island
hora, Czech, Slo. ... mountain
Horn, Ger. ... point, peak
hu, Ch. ... lake, reservoir
Hügel, Ger. ... hill
-huk, Swe. ... cape
ig., igarapé, Port. ... river
île(s), Fr. ... island(s)
îlet(s), Fr. ... islet(s)
ilha(s), Port. ... island(s)
ilhéu(s), Port. ... islet(s)
Insel(n), Ger. ... island(s)
isla(s), Sp. ... island(s)
isola, It. ... island
isole, It. ... islands
istmo, Sp. ... isthmus
jabal, Ara. ... mountain(s)
järv, Est. ... lake
-järvi, Finn. ... lake
jazā'ir, Ara. ... islands
jazirah, Indon. ... peninsula
jiang, Ch. ... river
-jima, Jpn. ... island
-joki, Finn. ... river
-jökull, Ice. ... glacier
-kai, Jpn. ... sea
-kaikyō, Jpn. ... strait
-kaise, Lapp. ... mountain
kali, Indon. ... brook
kandao, Pas. ... pass
-kang, Kor. ... river
-kapp, Nor. ... cape
kepulauan, Indon. ... islands
khalīj, Ara. ... gulf
khrebet, Russ., Ukr. ... range
-ko, Jpn. ... lake, lagoon
-kō, Jpn. ... harbor, harbour
kólpos, Gr. ... bay
Kopf, Ger. ... peak
körfezi, Tur. ... gulf, bay
kosa, Rus., Ukr. ... spit
kou, Ch. ... bay, pass
kuala, Mal. ... bay
kūh(ha), Per. ... mountain(s)
la, Tib. ... pass
lac(s), Fr. ... lake(s)
lag., laguna, Sp. ... lagoon, lake
lago, It., Port., Sp. ... lake
lagoa, Port. ... lake, lagoon
laguna, Sp. ... lagoon, lake
lagune, Fr. ... lagoon
laht, Est. ... bay
-lahti, Finn. ... gulf
län, Swe. ... county
laut, Indon. ... sea
liedao, Ch. ... islands
liman, Rus. ... estuary
ling, Ch. ... mountain(s), peak
llano(s), Sp. ... plain(s)
loch, Gae. ... lake, inlet
lomas, Sp. ... hills
lough, Gae. ... lake
lyman, Ukr. ... estuary
-maa, Est. ... island
-man, Kor. ... bay
mar, Sp., It. ... sea
marais, Fr. ... marsh
mare, It. ... sea
massif, Fr. ... massif
Meer, Ger. ... sea, lake
mer, Fr. ... sea
mesa, Sp. ... mesa
meseta, Sp. ... plateau
-misaki, Jpn. ... cape
mont, Fr. ... mount
montagna, It. ... mountain
montagne(s), Fr. ... mountain(s)
montaña(s), Sp. ... mountain(s)
monte, It., Port., Sp. ... mount
montes, Port., Sp. ... mountains
monti, It. ... mountains
monts, Fr. ... mountains
more, Rus., Ukr. ... sea
morne, Fr. ... mountain
morro, Port., Sp. ... hill, mountain
mui, Viet. ... point
munkhafad, Ara. ... depression
munţii, Rom. ... mountains
-nada, Jpn. ... sea, gulf
nafūd, Ara. ... desert
nagor'ye, Rus. ... plateau, mountains
-näs, Swe. ... peninsula
ness, Gae. ... promontory
nos, Blg. ... cape
nuruu, Mong. ... mountains
nuur, Mong. ... lake
-ø, Dan., Nor. ... island
-ö, Swe. ... island
o., ostrov, Rus. ... island
óros, Gr. ... mountain(s)
ostriv, Ukr. ... island
ostrov(a), Rus. ... island(s)
otok, S. \C. ... island
ouadi, Ara. ... wadi
oued, Ara. ... wadi
-øy, -øya, Nor. ... island
oz., ozero, Rus., Ukr. ... lake
pampa, Sp. ... plain
pas, Fr. ... strait
paso, Sp. ... pass
Pass, Ger. ... pass
passe, Fr. ... passage
passo, It. ... pass
peg., pegunungan, Indon. ... mountains
pélagos, Gr. ... sea
peña, Sp. ... peak, rock
península, Sp. ... peninsula
pertuis, Fr. ... strait
peski, Rus. ... sand desert
phnum, Cbd. ... mountain
phou, Lao. ... mountain
pic, Fr. ... peak
pico(s), Port., Sp. ... peak(s)
-piggen, Nor. ... mountain
pik, Rus. ... peak
pique, Fr. ... peak
piton(s), Fr. ... peak(s)
pivostriv, Ukr. ... peninsula
planalto, Port. ... plateau
planina, S. \C. ... mountain, range
plato, Afk., Blg., Rus. ... plateau
playa, Sp. ... beach
pointe, Fr. ... point
polje, S. \C. ... plain, basin
poluostrov, Rus. ... peninsula
pont, Fr. ... bridge
ponta, pontal, Port. ... point
porto, It. ... port
presa, Sp. ... reservoir, dam
presqu'ile, Fr. ... peninsula
proliv, Rus. ... strait
puerto, Sp. ... port
pulau, Indon., Mal. ... island
puncak, Indon. ... peak
punta, It., Sp. ... point, peak
qundao, Ch. ... islands
rāo., ribeirão, Port. ... river
ras, ra's, Ara. ... cape
rās, Per. ... cape
récif, Fr. ... reef
represa, Port. ... dam, reservoir
-retto, Jpn. ... islands
ría, Sp. ... ria (inlet)
rib., ribeira, Port. ... brook
ribeirão, Port. ... river
rio, Port. ... river
río, Sp. ... river
riviera, It. ... coast
rivière, Fr. ... river
roca, Sp. ... rock
rocca, It. ... rock, mountain
rt, S. \C. ... cape
sa., serra, Port. ... range
sahrā', Ara. ... desert
-saki, Jpn. ... cape
salar, Sp. ... salt flat
salina(s), Sp. ... salt marsh, salt flat
salto(s), Port., Sp. ... waterfall
-sammyaku, Jpn. ... range
-san, Jpn., Kor. ... mountain
-sanmaek, Kor. ... mountains
Schloss, Ger. ... castle
sebkha, Ara. ... salt flat
See(n), Ger. ... lake(s)
selat, Indon. ... strait
seno, Sp. ... sound
serra, Port. ... range, mountain
serranía(s), Sp. ... ridge(s)
shan, Ch. ... mountain(s), island
shanmo, Ch. ... mountains
-shima, Jpn. ... island
-shotō, Jpn. ... islands
sierra, Sp. ... range, ridge
-sjø, Nor. ... lake
-sjön, Swe. ... lake, bay
-sø, Dan. ... lake
Spitze, Ger. ... peak
sta., santa, Port., Sp. ... saint
ste., sainte, Fr. ... saint
step', Rus. ... steppe
štít, Slo. ... peak
sto., santo, Port., Sp. ... saint
stretto, It. ... strait
Strom, Ger. ... stream
-ström, -strömmen, Swe. ... stream
-su, Kor. ... river
-suidō, Jpn. ... channel
Sund, Ger. ... sound
-sund, Swe. ... sound
-take, Jpn. ... mountain
Tal, Ger. ... valley
tanjong, Mal. ... cape
tanjung, Indon. ... cape
tao, Ch. ... island
teluk, Indon. ... bay
thale, Thai ... lagoon
-tō, Jpn. ... island
tônlé, Cbd. ... lake
-tunturi, Finn. ... hill, mountain
ujung, Indon. ... cape
-umi, Jpn. ... lagoon
-ura, Jpn. ... lake
valle, It., Sp. ... valley
vallée, Fr. ... valley
vârful, Rom. ... mountain
-vatn, Ice., Nor. ... lake
vdkhr., vodokhranilishche, Rus. ... reservoir
-vesi, Finn. ... lake
-viken, Swe. ... gulf
vodokhranilishche, Rus. ... reservoir
vodoskhovyshche, Ukr. ... reservoir
vol., volcán, Sp. ... volcano
wādī, Ara. ... wadi
wāhat, wāhāt, Ara. ... oasis
wan, Ch., Jpn. ... bay
-yama, Jpn. ... mountain
yarımadası, Tur. ... peninsula
yoma, Bur. ... mountains
yumco, Tib. ... lake
yunhe, Ch. ... canal
-zaki, Jpn. ... point
zaliv, Rus. ... gulf, bay
zatoka, Ukr. ... gulf, bay
zee, Du. ... sea, lake

Abbreviations of Geographical Names and Terms

Abbreviation	Name	Abbreviation	Name
Ab., Can.	Alberta, Can.	Id., U.S.	Idaho, U.S.
Afg.	Afghanistan	Il., U.S.	Illinois, U.S.
Afr.	Africa	In., U.S.	Indiana, U.S.
Ak., U.S.	Alaska, U.S.	Indon.	Indonesia
Al., U.S.	Alabama, U.S.	Ire.	Ireland
Alb.	Albania	Is.	Islands
Alg.	Algeria	Isr.	Israel
Ang.	Angola	Jam.	Jamaica
Ant.	Antarctica	Jord.	Jordan
Ar., U.S.	Arkansas, U.S.	Kaz.	Kazakhstan
Arg.	Argentina	Ks., U.S.	Kansas, U.S.
Arm.	Armenia	Kuw.	Kuwait
Aus.	Austria	Ky., U.S.	Kentucky, U.S.
Austl.	Australia	Kyrg.	Kyrgyzstan
Az., U.S.	Arizona, U.S.	L.	Lake
Azer.	Azerbaijan	La., U.S.	Louisiana, U.S.
B.	Bay	Lat.	Latvia
Bah.	Bahamas	Leb.	Lebanon
Bahr.	Bahrain	Leso.	Lesotho
Barb.	Barbados	Lib.	Liberia
B.C., Can.	British Columbia, Can.	Lith.	Lithuania
Bdi.	Burundi	Lux.	Luxembourg
Bel.	Belgium	Ma., U.S.	Massachusetts, U.S.
Bela.	Belarus	Mac.	Macedonia
Bhu.	Bhutan	Madag.	Madagascar
Bngl.	Bangladesh	Malay.	Malaysia
Bol.	Bolivia	Mart.	Martinique
Bos.	Bosnia and Herzegovina	Maur.	Mauritania
Bots.	Botswana	Mb., Can.	Manitoba, Can.
Braz.	Brazil	Md., U.S.	Maryland, U.S.
Bul.	Bulgaria	Me., U.S.	Maine, U.S.
Burkina	Burkina Faso	Mex.	Mexico
C.	Cape	Mi., U.S.	Michigan, U.S.
Ca., U.S.	California, U.S.	Mn., U.S.	Minnesota, U.S.
Camb.	Cambodia	Mo., U.S.	Missouri, U.S.
Can.	Canada	Mol.	Moldova
C.A.R.	Central African Republic	Mong.	Mongolia
Cay. Is.	Cayman Islands	Monts.	Montserrat
C. Iv.	Cote d'Ivoire	Mor.	Morocco
Co., U.S.	Colorado, U.S.	Moz.	Mozambique
Col.	Colombia	Ms., U.S.	Mississippi, U.S.
C.R.	Costa Rica	Mt.	Mountain
Cro.	Croatia	Mt., U.S.	Montana, U.S.
Ct., U.S.	Connecticut, U.S.	Mts.	Mountains
Ctry.	Country	Mwi.	Malawi
C.V.	Cape Verde	Myan.	Myanmar
Cyp.	Cyprus	N.A.	North America
Czech Rep.	Czech Republic	N.B., Can.	New Brunswick, Can.
D.C., U.S.	District of Columbia, U.S.	N.C., U.S.	North Carolina, U.S.
De., U.S.	Delaware, U.S.	N. Cal.	New Caledonia
Den.	Denmark	N.D., U.S.	North Dakota, U.S.
Dep.	Dependency	Ne., U.S.	Nebraska, U.S.
Des.	Desert	Neth.	Netherlands
Dji.	Djibouti	Neth. Ant.	Netherlands Antilles
D.R.C.	Democratic Republic of the Congo	N.H., U.S.	New Hampshire, U.S.
Ec.	Ecuador	Nic.	Nicaragua
El Sal.	El Salvador	Nig.	Nigeria
Eng., U.K.	England, U.K.	N. Ire., U.K.	Northern Ireland, U.K.
Eq. Gui.	Equatorial Guinea	N.J., U.S.	New Jersey, U.S.
Erit.	Eritrea	N. Kor.	North Korea
Est.	Estonia	N.L., Can.	Newfoundland and Labrador, Can.
Eth.	Ethiopia	N.M., U.S.	New Mexico, U.S.
Eur.	Europe	Nmb.	Namibia
Falk. Is.	Falkland Islands	Nor.	Norway
Fin.	Finland	N.S., Can.	Nova Scotia, Can.
Fl., U.S.	Florida, U.S.	N.T., Can.	Northwest Territories, Can.
Fr.	France	Nu., Can.	Nunavut, Can.
Fr. Gu.	French Guiana	Nv., U.S.	Nevada, U.S.
G.	Gulf	N.Y., U.S.	New York, U.S.
Ga., U.S.	Georgia, U.S.	N.Z.	New Zealand
Gam.	The Gambia	Oc.	Oceania
Gaza Str.	Gaza Strip	Oh., U.S.	Ohio, U.S.
Geor.	Georgia	Ok., U.S.	Oklahoma, U.S.
Ger.	Germany	On., Can.	Ontario, Can.
Grc.	Greece	Or., U.S.	Oregon, U.S.
Guad.	Guadeloupe	Pa., U.S.	Pennsylvania, U.S.
Guat.	Guatemala	Pak.	Pakistan
Gui.	Guinea	Pan.	Panama
Gui.-B.	Guinea-Bissau	Pap. N. Gui.	Papua New Guinea
Guy.	Guyana	Para.	Paraguay
Hi., U.S.	Hawaii, U.S.	P.E., Can.	Prince Edward Island, Can.
Hond.	Honduras	Pen.	Peninsula
Hung.	Hungary	Phil.	Philippines
I.	Island	Pk.	Peak
Ia., U.S.	Iowa, U.S.	Plat.	Plateau
Ice.	Iceland	Pol.	Poland
		Port.	Portugal
		P.R.	Puerto Rico
		Prov.	Province
		Qc., Can.	Quebec, Can.
		R.	River
		Ra.	Range
		Reg.	Region
		Res.	Reservoir
		R.I., U.S.	Rhode Island, U.S.
		Rom.	Romania
		Rw.	Rwanda
		S.A.	South America
		S. Afr.	South Africa
		Sau. Ar.	Saudi Arabia
		S.C., U.S.	South Carolina, U.S.
		Scot., U.K.	Scotland, U.K.
		S.D., U.S.	South Dakota, U.S.
		Sen.	Senegal
		Serb.	Serbia and Montenegro
		Sk., Can.	Saskatchewan, Can.
		S. Kor.	South Korea
		S.L.	Sierra Leone
		Slvk.	Slovakia
		Slvn.	Slovenia
		Som.	Somalia
		Sp. N. Afr.	Spanish North Africa
		Sri L.	Sri Lanka
		St. Vin.	St. Vincent and the Grenadines
		Sur.	Suriname
		Swaz.	Swaziland
		Swe.	Sweden
		Switz.	Switzerland
		Tai.	Taiwan
		Taj.	Tajikistan
		Tan.	Tanzania
		Ter.	Territory
		Thai.	Thailand
		Tn., U.S.	Tennessee, U.S.
		Trin.	Trinidad and Tobago
		Tun.	Tunisia
		Tur.	Turkey
		Turk.	Turkmenistan
		Tx., U.S.	Texas, U.S.
		U.A.E.	United Arab Emirates
		Ug.	Uganda
		U.K.	United Kingdom
		Ukr.	Ukraine
		Ur.	Uruguay
		U.S.	United States
		Ut., U.S.	Utah, U.S.
		Uzb.	Uzbekistan
		Va., U.S.	Virginia, U.S.
		Ven.	Venezuela
		Viet.	Vietnam
		V.I.U.S.	Virgin Islands (U.S.)
		Vol.	Volcano
		Vt., U.S.	Vermont, U.S.
		Wa., U.S.	Washington, U.S.
		Wal./F.	Wallis and Futuna
		W.B.	West Bank
		Wi., U.S.	Wisconsin, U.S.
		W. Sah.	Western Sahara
		W.V., U.S.	West Virginia, U.S.
		Wy., U.S.	Wyoming, U.S.
		Yk., Can.	Yukon Territory, Can.
		Zam.	Zambia
		Zimb.	Zimbabwe

Index

This universal index includes in a single alphabetical list approximately 4,100 names of features that appear on the reference maps. Each name is followed by geographical coordinates and a page reference.

Abbreviation and Capitalization

Abbreviations of names on the maps have been standardized as much as possible. Names that are abbreviated on the maps are generally spelled out in full in the index. Periods are used after all abbreviations regardless of local practice. The abbreviation "St." is used only for "Saint". "Sankt" and other forms of this term are spelled out."

Most initial letters of names are capitalized, except for a few Dutch names, such as "'s-Gravenhage." Capitalization of non-initial words in a name generally follows local practice.

Alphabetization

Names are alphabetized in the order of the letters of the English alphabet. Spanish ll and ch, for example, are not treated as distinct letters.

Furthermore, diacritical marks are disregarded in alphabetization. German or Scandinavian ä or ö are treated as a or o.

The names of physical features may appear inverted, since they are always alphabetized under the proper, not the generic, part of the name, thus: "Gibraltar, Strait of." Otherwise every entry, whether consisting of one word or more, is alphabetized as a single continuous entity. "Lakeland," for example, appears after "Lake Forest" and before "La Línea." Names beginning with articles (Le Havre, Al Manāmah, Ad Dawhah) are not inverted. Names beginning "St.," "Ste.'" and "Sainte" are alphabetized as though spelled "Saint."

In the case of identical names, towns are listed first, then political divisions, then physical features.

Generic Terms

Except for cities, the names of all features are followed by terms that represent broad classes of features, for example, *Mississippi, R.* or *Alabama, State.*

Country names and names of features that extend beyond the boundaries of one country are followed by the name of the continent in which each is located. Country designations follow the names of all other places in the index. The locations of places in the United States, Canada, and the United Kingdom are further defined by abbreviations that indicate the state, province, or political division in which each is located.

Page References and Geographical Coordinates

The geographical coordinates and page references are found in the last columns of each entry.

Latitude and longitude coordinates for point features, such as cities and mountain peaks, indicate the locations of the symbols. For extensive areal features, such as countries or mountain ranges, or linear features, such as canals and rivers, locations are given for the position of the type as it appears on the map.

Index